D0712609

Colored Canaries

Colored Canaries

G. B. R. Walker

Photographs and illustrations by

Dennis Avon & Tony Tilford

ARCO PUBLISHING COMPANY, INC.
New York

This book is dedicated to Ruth, Maria and Elizabeth without whose remarkable forbearance nothing would have been possible.

GBRW

Published 1977 by Arco Publishing Company, Inc.
219 Park Avenue South, New York, N.Y. 10003

Printed in Great Britain

Library of Congress Cataloging in Publication Data

Walker, G B R
Colored canaries.

Arco color series
Includes index.
1. Color canaries. 2. Canaries—Breeding.
I. Title.
SF463.7.C64W34 635.6'862 76–51796
ISBN 0 668 04207 9

Contents

Acknowledgements

The author particularly wishes to thank Bob Yates, Mike Attew, Jim Swarbrick and Mick Watton for all their help in checking the text and photographs used in this book; and to Monsieur Ascheri of Paris, for the considerable technical assistance given and for writing the Foreword to this book. To Alice for the time given in typing the manuscript and all members of the Canary Colour Breeders Association who have been so free in giving their advice and suggestions, and for allowing their birds to be photographed.

The colour plates were prepared from photographs of birds kindly loaned by the following people:

M. Ascheri, M. Attew, A. Benbow, J. Bradshaw, M. Chisholme, Mr and Mrs E. Clibbon, J. Cowell, R. Dexter, S. Edwards, T. Halford, P. Harrison, G. Izzo, A. Jones, Mrs C. Lewis, S. Orton, L. Roberts, L. Rooker, C. Sterry, N. Thompson, G. Walker, B. Warren, G. Watson, T. Watson, M. Watton, S. West, P. Williams.

List of Colour Plates

Foreword

The book which I have the pleasure of introducing is, I feel, of particular distinction as I do not believe that any author has ever combined in one volume such essential information with such remarkable photographs. Only with the patience and undoubted talent of a man like Geoff Walker could such a work have been accomplished.

The author is a young but very experienced breeder whom I first had the pleasure of meeting at the British National Exhibition of Cage and Aviary Birds at Alexandra Palace, London several years ago and it is an honour for me to have been asked to supply the Foreword to *Coloured Canaries*.

Above all, the book will help amateur breeders to discover and differentiate the varieties of coloured canaries. The photographers, Dennis Avon and Tony Tilford, took several hundred photographs before selecting the very best ones to illustrate this book. You may already have seen some of their photographs in *Aviary Birds in Colour* and *Birds of Britain and Europe in Colour* and will therefore appreciate their superb work.

Mr Walker has spent much of his time in breeding research in order to enlarge his already considerable experience as a breeder and judge – experience which is complemented by a great love of birds and a dedication to aviculture.

Much of his time and energy was devoted to cross-breeding. Diversifying and expanding his techniques, he tested with precision the laws of genetics which govern the breeding of canaries.

In the breeding of canaries, the male inherits his melanistic coloration from both the mother and the father; but the female only from the father. The explanation for this is that only the heterozygote chromosome of the female contains the sex-linked genetic factors; the other chromosome Y being empty. The male canary carries the colour characteristics of both X and X sex chromosomes.

As with Mendel and the peas, Blakeslee with the *Datura* and Morgan with the *Vinagre drosophila*, Geoff Walker is meticulous and persistent in his breeding techniques for colour canaries. He excels in retracing the history of the canary from the first mating of a male Hooded Siskin with the recessive yellow and white female canary.

It was a great honour to welcome Geoff Walker, together with Dennis Avon and Tony Tilford, to my Centre Technique D'Elevage in Paris. They were able to observe, photograph and later describe in detail my latest results. Several of these mutations only recently fixed will I hope provide keen amateur breeders with new areas to investigate with their own canaries.

M. Ascheri

Centre Technique d'Elevage du Canari et de ses Hybrides
72 rue du Rendezvous
Paris

August 1976

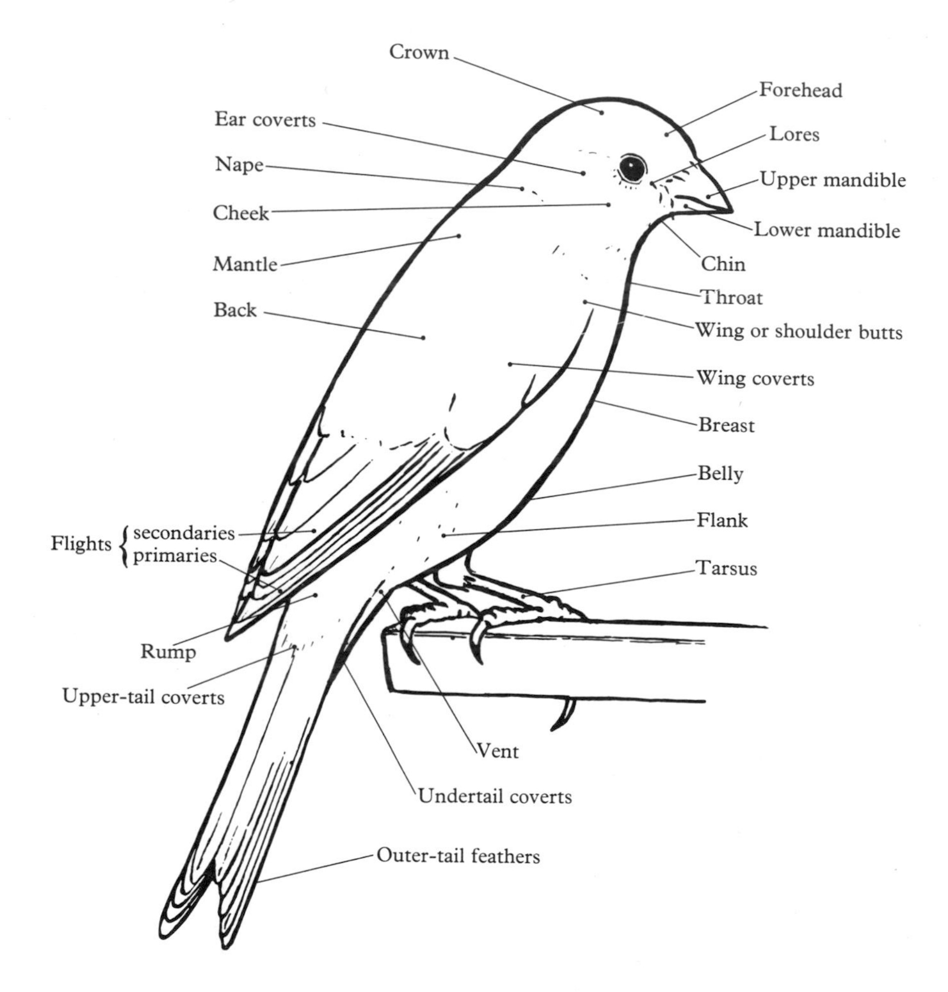

Topography of the Canary

Introduction and History

'The longing for something to protect and care for is one of the strongest feelings implanted within us, and one outcome of it is the desire to keep animals and birds under our control, which in its due place is undoubtedly, one of our healthiest instincts.' So says W. A. Blakston, in his book *The Illustrated Book of Canaries and Cage Birds*, published by Cassell and Company, at about the turn of the century. Today, for many of us, the same feelings apply. Couple with this the increasing need for a stimulating relaxing hobby to help combat the ever-demanding pace set by today's world it is obvious why the keeping, breeding and exhibiting of domesticated cage birds in general, and canaries in particular, is becoming more and more popular.

The original domestication of the wild Canary (*Serinus canarius*) appears to have been carried out in or about the fifteenth century by the Spaniards, who trapped and caged the birds for their song. At this point two theories on their distribution arise. The first, is that Canaries were shipped on a fairly regular basis to Europe from that time. Initially, only males were marketed by the shrewd Spaniards, thus effectively retaining a monopoly on sales. Eventually, however, a supply of females became available and the breeding of Canaries in captivity swiftly spread throughout Europe.

The second, is that a consignment of Canaries were on board a ship that was wrecked off the island of Elba. The birds were initially set free but, after a short period, were recaptured and shipped to Italy as song birds. Eventually the birds found their way into Germany and from then the two stories start to coincide.

By the seventeenth century records show that canaries were being bred in Spain, Italy, France, Switzerland and southern Germany. It was in the last-named country that the Hartz Mountain canary, renowned for its song was evolved.

In its early history of domestication the black and brown melanins were bred out, leaving a clear yellow bird. As it continued its path across Europe it became more a bird of posture than of song. This was particularly true in Britain, the tradition of excellence in posture canaries being carried through to modern times. British posture or type canaries are still greatly sought throughout the world. The wild Canary, according to current hypotheses, has four colour pigments, one black, two brown, and one yellow. The black is located down the centre of the feather and on the underfeather and is known as 'eumelanin black'. The first of the brown pigments is located round the edge of the feather and is known as 'phaeomelanin brown'. The second of the brown pigments – eumelanin brown – is located down the centre of the feather mingled with the black. These pigments are called 'melanins' and are formed from protein produced by the birds. The yellow is a fat-soluble colouring material known as 'lipochrome' and belongs to the class of carotenoid pigments found in plants, and is derived from the food which the bird eats.

At some time, and it is not known whether this was in the wild or during its domestication, a spontaneous change occurred on the gene producing the eumelanin black, which altered the colour throughout the feather from black to brown. This spontaneous change, or mutation, resulted in a brown-coloured bird with a pink eye, which was then known as 'cinnamon' but is known today as 'brown'. The clear yellow canary although devoid of melanin is still genetically a normal or green, and consequently it is possible to breed a clear yellow canary that is genetically a brown. The only indication of this would be in the colour of the eye in the nest.

Other spontaneous occurrences may have happened, but were either not recognised or were lost, and the only other mutation to have survived the nineteenth century is thought to be the dominant white. It is not known if the bird we call the dominant white, which became prevalent about 1920, is the same mutation as the white birds recorded much earlier. The effect of the dominant white was to mask almost entirely the yellow lipochrome present in both clear and self canaries. Thus at the turn of the century there existed clear yellow and white canaries, yellow canaries with the brown and black melanins (greens), white canaries with the brown and black melanins (blues), yellow canaries with the brown and mutated brown melanins (cinnamons, now known as 'gold browns') and white canaries with the brown and mutated brown melanins (fawns, now known as 'silver browns').

No sooner had the twentieth century seen the light of day, than a further mutation occurred. In a Dutch breeder's stud, a pair of green canaries produced an ash-grey offspring. Close examination showed that all the melanins had their intensity diluted. The bird was called an 'agate' and test matings showed that the mutation was sex linked. With the help of a crossover of genes from one of the two sex chromosomes to another, it was comparatively easy to transfer the factor to a brown bird. The combination of the brown factor and the new diluting factor was called 'isabel' – and another new colour had made its début.

The next mutation to appear that made any lasting impression was the second of the white factors. Unlike its dominant counterpart the second white factor was recessive to normal, being visible only when present on both lipochrome colouring genes. This type made its appearance about 1908, and was found to be different from the dominant white canary in that it completely masked all traces of lipochrome.

Man, impulsive and creative animal that he is, tired of waiting for another mutation to occur and thought was given to attempting to produce a red canary. Dr Duncker, a leading fancier of the day from Germany, put forward the theory that the red-producing gene could be transferred to the canary by hybridisation. The bird chosen for these experiments was the Black-hooded Red Siskin (*Spinus cucullatus*). This small vermilion-coloured bird with black hood and wings, a native of Venezuela and the Monas Islands, proved a ready subject for hybridisation, and limited fertility was found in the first cross siskin × canary males. These birds are known as 'F1 hybrids'. It was not until the second outcross that fertile females were produced, and the quest for the red canary could begin in earnest. Eventually the gene for red was more or less stabilised in the canary, when the experimental breeders of the day proceeded not only to improve the coloration in the clear red birds, but also to transfer the red gene to the existing self-mutations.

In 1949 a further mutation occurred in Germany, but it was not until the 1960s that it made any impression on the coloured canary breeders. This mutation which prevented the phaeomelanin brown from expressing itself and which changed the colour of the eumelanin black to a silvery grey colour, was labelled the 'opal'.

Prior to the opal factor gaining popularity another mutation occurred in a strain of clear red birds. A Dutch breeder produced in 1951 a pale coloured youngster from a normal red orange × apricot pairing. This youngster when moulted showed that the normal red coloration had been modified to a rose pink. The factor proved to be sex linked, and it is only in the last three or four years that we have seen it introduced to any degree of perfection in the self series. There are birds showing this characteristic in all ground colours, clear, variegated and self. Yellow ground birds are described as gold ivory, white ground as silver ivory and orange ground as rose.

In the late 1950s, also in a Dutch breeder's room, the conception of certain dedicated fanciers who had been attempting to breed an isabel canary devoid of melanins came to fruition. This was caused, not by their efforts but by the appearance of a further mutation – the pastel. The full effect of this variety is still subject to a great deal of debate, but it would appear to modify the intensity of eumelanin brown, which when accompanied by the isabel factor causes this pigmentation to disappear completely. The effect on the green series is more complex giving a variety of different appearances. We will examine these more closely later in this book.

Two other mutations have occurred since the pastel, one a sex-linked recessive and the other a homozygous recessive factor. Both are totally different from any of their predecessors in that they result in a bird with distinctive clear red eyes, which, unlike the eyes of the brown canary, do not fade as the chick grows. The effect on the melanins is, however, totally different. The first of the two to appear – the double recessive, known as the 'ino' – obliterates all trace of eumelanin black, and modifies to a small extent the eumelanin brown. The phaeomelanin brown is not affected and gives the brown version in particular a hammered copper effect. The 'satinette' factor, as the sex-linked version is known, appeared in the late 1960s, and is now only beginning to arrive in any numbers on the showbench. Being comparatively new, we obviously have not exhausted all possible effects it may have, but we are fairly sure that the only melanin pigment the factor does not affect is eumelanin brown.

This then is a potted history of coloured canaries up to the present time. This section of the fancy is still very much in its infancy, and as such incorporates so many different points of view that all of them must remain partially hypothetical. As time moves on and more and more people start breeding these fascinating birds, it is hoped that these notes will provide a basis for healthy discussion so that together we can truly unravel more of the mysteries that now surround our multi-coloured friends.

Inheritance

In order that we, as practical breeders of canaries, can achieve our objectives, it is necessary for us to have some knowledge of the laws of heredity. Canaries being low in priority where the betterment of man's position is concerned, few truly scientific projects have been undertaken to unravel the mysteries surrounding our subject and much still remains to be ascertained. However elementary our knowledge, if we understand it more fully it will serve us twofold. Firstly, the chances of us achieving set targets will improve if we adopt and follow certain natural laws and, secondly, we will be more capable of recognising and exploiting as yet unexplained phenomena.

The word 'genetics' has in its time instilled more fear into practical breeders than any other, but we all rely for success on the laws of heredity whether consciously or not.

Gregor Mendel, an Austrian monk, who died in 1884, is the father of the modern science of genetics. As his subject he chose pea plants. He had noticed that 'tall' peas always produced tall pea plants, while 'dwarf' peas always produced dwarf pea plants, and he contemplated the result of cross-breeding the two. He found the first cross hybrids (F1 or first filial generation) were all tall plants, not medium sized as might have been expected. This character Mendel called 'dominant' and the dwarf character which did not appear he called 'recessive'. On self-fertilising these F1 hybrids, the progeny, called 'F2' or second filial generation, appeared in the proportion of three tall to one dwarf. The tests on these plants showed that the dwarf plants always bred true, but that only one in three of the tall plants bred true. The other two plants produced both tall and dwarf plants, Mendel called the dwarf ones 'pure recessives', and the true breeding tall ones 'pure dominants'. The two other tall plants, although identical in appearance to their true breeding counterparts, were hybrids and he called them 'impure dominants'.

In all experiments, when a pair of different contrasting characters were involved where one was dominant to the other, three types of plant appeared:

1 Pure dominants.
2 Impure dominants.
3 Pure recessives.

Further experiments proved that the pure recessives and pure dominants always bred true, while the impure dominants produced pure dominants, impure dominants and pure recessives. These on average appeared in the proportion of 1:2:1.

All creation starts with the germ cells called 'gametes'. A gamete from the male joins with a gamete from the female (fertilisation) and forms a new cell which is termed the 'zygote'. It is in this cell that all the potentialities of new organism are found.

Mendel used a capital A to represent the dominant tall factor which was found in the germ cells produced from a true breeding tall and the small a to represent the recessive dwarf factor produced from the gametes of a true breeding dwarf. Thus the true breeding tall was represented by the figures AA, having received one A factor from each of its parents. The dwarf was designated aa having received an a factor from each of its parents.

The F1 hybrid obtained by crossing these two species received one A factor and one a factor, and was thus represented as Aa. As all the plants were tall the capital A was seen to be dominant over the recessive a. On forming gametes itself the hybrid produced two different types. The first contained the A factor and the second the a factor. This proved that the factors, although contrasting, instead of uniting in the cells of the hybrid remained separated and on the formation of the gametes segregated independently.

On self-fertilisation the A of the male gamete could join with either the female A gamete or the female a gamete. From the chart on opposite page we can see that the only possibilities open are the production of two types of plant. (1) AA being a true breeding tall and (2) Aa being visually a tall, but capable of producing a dwarf.

Similarly a male a gamete can join with either a female A gamete or a female a gamete. This gives results of either an aA offspring, being a tall incapable of breeding true or an aa which is a true breeding dwarf.

Thus on average the results of the association of four factors must be one AA (pure breeding tall). Two Aa (or aA) visual talls but carrying the factor for dwarfness and one aa (a pure breeding dwarf). This returns us to the Mendelian ratio on which we have commented 1:2:1.

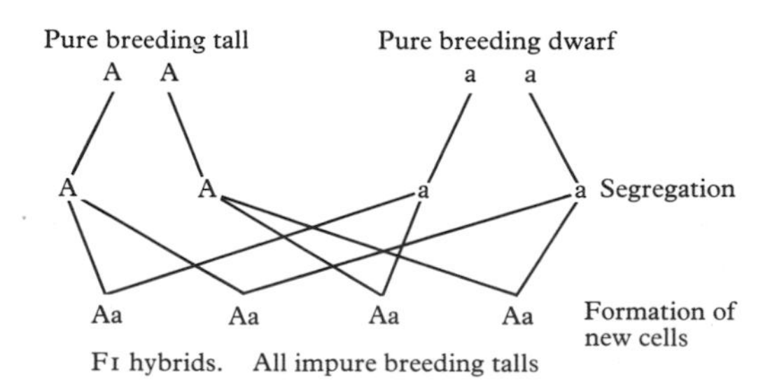

F1 hybrids. All impure breeding talls

Thus the two fundamental laws of Mendel are, firstly the segregation of the gametes which result in the 1:2:1 ratio in the second filial generation and, secondly, the law that shows the various factors concerned segregate independently of each other.

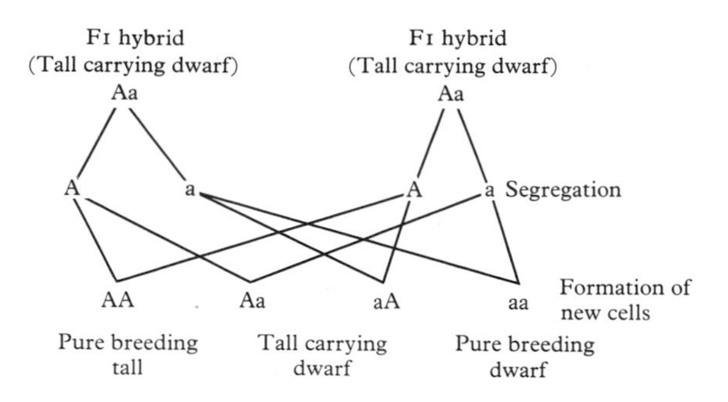

These factors in relation to heredity in plants are equally relevant when applied to the canary.

A canary chick comes from an egg (ovum) produced by a female canary which has been fertilised with sperm produced by a male canary. Within the sperm and egg (germ cells) the most obvious thing is the nucleus. This contains a fixed number of threadlike units (chromosomes) on which are located genes or factors which control the hereditary characteristics. These genes appear as beads on a string – the chromosomes. The germ cells differ from the body cells, known as the 'somatic cells', in that they have only one each of the pairs of chromosomes that exist in the body cells. In all bisexual species both the male and the female have similar chromosomes, with the exception of the sex chromosome. In canaries, the male sex chromosomes are alike, and are designated XX. The females are not alike and are designated XY. The Y chromosomes carries only genes that determine femaleness.

Before a male or female germ cell is formed the parent cell from which it is derived has the full number of chromosomes in pairs all bunched together. When this cell divides to form two ova or sperms the chromosomes unravel and one from each pair goes to one side with its partner going to the other side. Thus when the sperms and ova fuse a full complement is again formed.

The movement of the chromosomes on separation is completely at random as can be seen by examining the inheritance pattern of the sex chromosomes. On average there should be equal numbers of males and females hatched, as half the ova have an X chromosome and half have a Y chromosome. As we all know, however, it is not uncommon to produce all males or all females in a nest, this being due to the action of the laws of chance.

There are many cells within a bird, each of which is different and has a different function. Within each cell is a nucleus which contains the chromosomes. Located on the chromosomes are the genes. The genes are the agents responsible for carrying out different processes within the cell. When the like or corresponding gene on each chromosome of the parents is identical, the resultant linked pair is then known as 'homozygous'; when different the resultant pair is said to be 'heterozygous'. A heterozygous bird is often referred to as being 'cross-bred'. It should be noted that a bird can be homozygous in many ways, but heterozygous in others, e.g. an agate male carrying isabel is homozygous for the agate factor, i.e. both genes are identical, but is also heterozygous in that only one of its X chromosomes is carrying the mutated brown gene.

We now use these terms instead of the Mendelian phraseology, so that a pure dominant becomes a homozygous dominant, an impure dominant becomes a heterozygous dominant and a pure recessive becomes a homozygous recessive.

When the heterozygous dominant has the same appearance as the homozygous dominant the birds are said to have the same 'phenotype', but as their genetical make-up is different they belong to different genotypes.

Occasionally accidents occur that can affect chromosomes or genes. Two forms of accident are known, the first is called a 'crossover' and will be studied later. The second is called a 'mutation', which is a spontaneous reaction changing the normal functions of a gene. Mutations occur only in very rare instances and cannot be made to happen although exposure to radiation has been known to increase the frequency in certain spheres. When mutated, a gene does not alter its position on the chromosome, but produces a different effect. Most mutations have some obscure effect, and, when mutated, a gene can be more or less effective than its unchanged partner. Most mutated genes are less effective than their unchanged partner in which case they are said to be 'recessive'. In certain instances, however, they are more effective and are then said to be 'dominant'. Once a mutation has taken place the gene obviously exists in two forms, the original one which produces a known and familiar effect and the mutant form which produces a new effect. Two different forms of the same gene are called 'allelomorphs', the original one being known as the 'wild allelomorph' or 'normal allelomorph', and the new one as the 'mutant allelomorph'.

Before proceeding with a consideration of the inheritance pattern of mutated genes, we should first examine the hereditary pattern of the sex chromosomes. This will become more relevant at a later stage.

We know that the male bird has two sex chromosomes on which genes are present and is designated XX, while the female has only one. She is designated XY.

We also know that these chromosomes split when germ cells are being produced and link with a chromosome from the other parent to form a new being. One of the X chromosomes of the male bird can, on fusion, join with either the X or the Y chromosome of the female. If it joins with the X chromosome a young male bird is produced, if with the Y chromosome a young female results. Using the chart below we can see that the determination of sex should in theory be 50 per cent males and 50 per cent females.

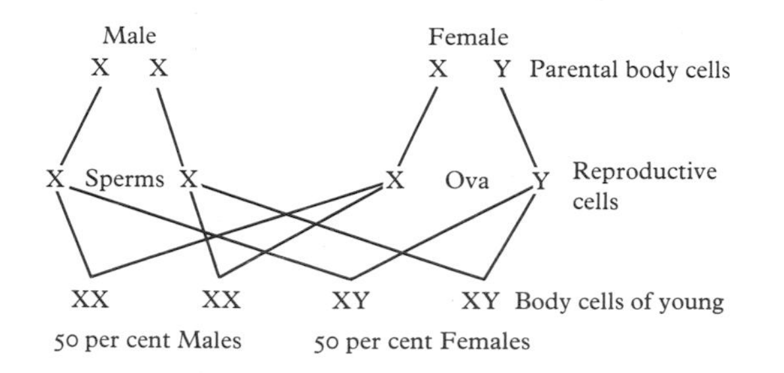

We can now see that the male canary is unable to determine the sex of the young. The female produces a certain number of eggs which will produce only young males (X chromosome) and a certain number that will produce only young females (Y chromosome).

A gene situated on the sex chromosome may mutate, and when this happens its inheritance pattern is dictated by the sex of the young produced. This is referred to as a 'sex-linked mutation', and to date all such mutations have proved to be recessive to the normal unchanged gene. The first of these mutations to appear in the canary is thought to be the brown (cinnamon). To be visible the mutation has to be present in a double dose, unless the recipient is a female in which case, if the single gene she possesses on her X chromosome mutates, she will be visually brown. Let us assume that we possess one such female bird, and we pair her to a green male bird following the same chart that we used for the determination of sex. We will use the letter B to indicate that the gene has mutated to produce brown.

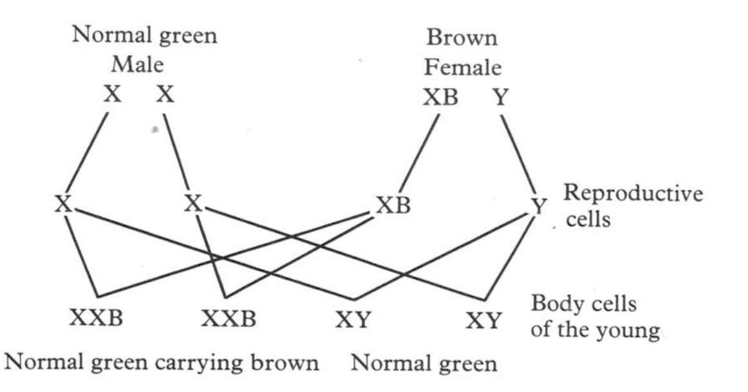

From this we can see that B (the mutated gene) is now present on one of each pair of chromosomes received by every male bird produced, yet none of the young females carry the mutated gene. The young males will be the same phenotype as their father but have a different genotype. The sex chromosomes will now be heterozygous and they will be capable of producing both green and brown females, although they can produce only homozygous green males and heterozygous green males carrying the factor for brown, viz.

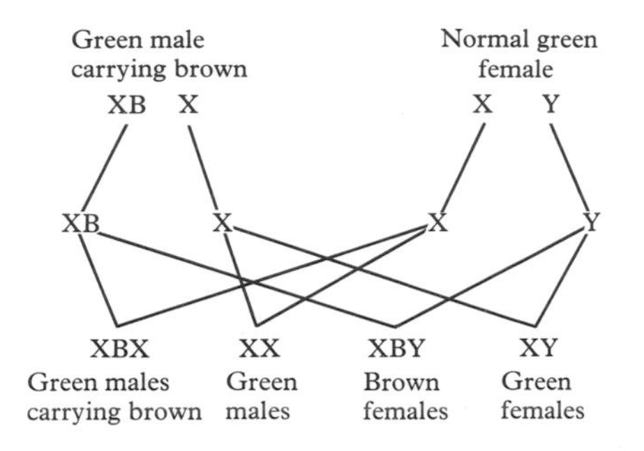

By backcrossing the green male carrying brown to its grandmother we can illustrate the progression of the hereditary pattern. The mutated gene of the male can join with the mutated gene of its grandmother to produce brown males, or it can pair with the Y chromosome to produce brown females. Conversely the unchanged gene can join with its grandmother's mutated gene to produce green males carrying the factor for brown. If it joins with the Y chromosome of its grandmother, green females will be produced.

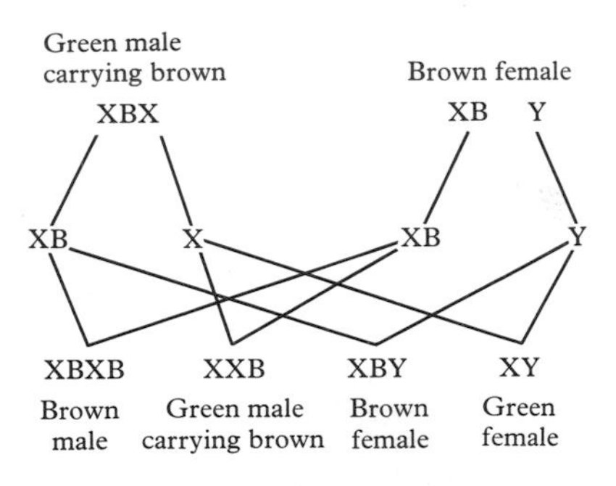

In addition to the brown, there exist today other sex-linked factors which are of particular interest to the breeder of coloured canaries. These are the isabel, agate, pastel and satinette factors which, apart from eye colour visually affect only those birds displaying melanistic pigment, and also the ivory factor which affects the lipochrome coloration of the bird (ground colour). In each instance the same rules apply and by selecting another letter or symbol it is easy to form an inheritance and expectancy table to cover all eventualities.

Two other forms of mutation may occur within the genes of the somatic chromosomes. In some cases the phenotype of the bird alters when one of a pair of genes mutates, these are known as 'heterozygous dominant mutations'. When it is necessary for both of the pair of genes to have mutated before the bird is visually different, the mutation is said to be a 'homozygous recessive mutation'.

Among the former are the non-frosted feather pattern, and the dominant white mutation, while among the latter we can list the ino and opal mutations.

We will take each type of mutation in turn using the dominant white as an example to illustrate the former. As its name implies, the white of the dominant white is dominant over the normal yellow-producing gene. Thus a single dominant white gene is sufficient to produce an almost white canary. The statement is qualified by the word 'almost', as in this and other instances, while the mutated gene is so much more effective than its unchanged partner, it is not totally dominant over the unchanged gene. In the case of the dominant white bird, certain areas will in most examples still show as yellow. It would be totally impracticable to attempt to qualify each factor as partially dominant or recessive and hence if the alteration in the normal pattern is sufficiently obvious for there to be no doubt that a mutation has taken place, it is referred to as dominant.

As the dominant white gene is not located on the sex chromosome we cannot know whether it will reappear in the males or in the young females. The mother possesses two gene-carrying chromosomes and either of these may mutate and be passed to either a young male or a young female.

By studying the possible matings to produce dominant whites we can illustrate how the Mendelian 1:2:1 theory is relative to canaries.

1 Dominant white male × yellow female
  Result: 50 per cent white young : 50 per cent yellow young
2 Yellow male × dominant white female
  Result: 50 per cent white young : 50 per cent yellow young
3 Dominant white male × dominant white female.
  Result: 50 per cent dominant white young (heterozygous)
  25 per cent dominant white young (homozygous)
  25 per cent yellows

To see how the results of the third pairing occur study the chart overleaf where a capital W is used to denote the mutated gene. A small w indicates the normal yellow producing gene.

We can see that 25% are WW being 000 homozygous dominant whites (these birds are actually non-viable and will not live, a subject we will examine further when the mutation is studied more closely).

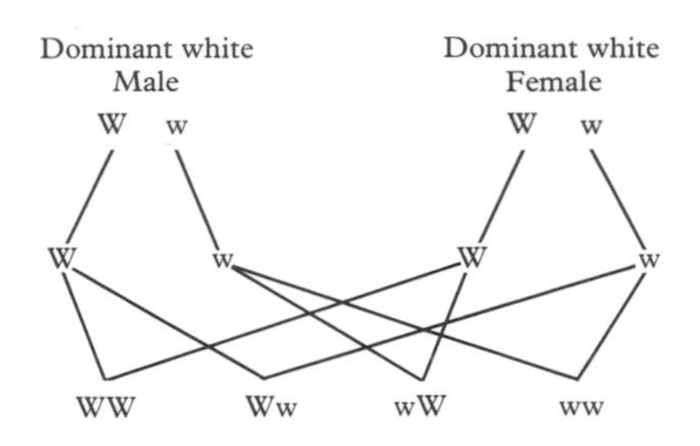

Fifty per cent are either Ww or wW (which is the same) these are heterozygous dominant whites and twenty-five per cent homozygous normal yellows – a result which conforms with the Mendelian ratio 1:2:1.

The action of a homozygous recessive mutation follows a similar pattern to that of the dominant white but it is necessary for both genes of a pair to have mutated before the mutation becomes visible. To illustrate this we will take as an example the opal mutation using the capital O to indicate the mutated gene, and the small o as the normal unchanged gene. When the capital O appears in double dose the effect of the opal mutation will be visible. As with the dominant white the gene that has mutated is not on the sex chromosome so we shall be unable to distinguish between the young males and females. Taking initially a visual opal male paired to a normal female the pairing would be represented as OO × oo which on separation and rejoining could only produce Oo in any of the four combinations. This bird will be identical in appearance to its mother, but will carry the factor for opal. If we pair brother to sister from the progeny we have Oo × Oo which could produce 25 per cent OO, 50 per cent Oo and 25 per cent oo, or 25 per cent opals, 50 per cent normals carrying the opal factor and 25 per cent homozygous normals. Again the heterozygous and the homozygous normals will have the same appearance, and only by test mating these birds, that is by mating them with a bird of known genotype, i.e. a full opal, could we determine which is which.

Nowadays it is comparatively easy for us to acquire such birds so that we can make the test matings, but please consider the problems and patience of the early breeders who were faced with the task of establishing such a mutation when often all they had to work with were birds that possibly carried the factor.

We must return to the classic example of attempting to produce a dilute green (agate) from dilute cinnamon (isabel) parents to demonstrate the occurrence of a crossover. From breeding experiences we know that this is possible but if we make a chart similar to those used in the determination of sex and inheritance patterns of the brown mutation, it would seem impossible. For the first time we are considering the inheritance patterns of two sex-linked mutations at the same time. These being mutations of different genes on the same chromosome. We will continue to use the symbol B to denote brown, and will introduce D to indicate dilute (agate or isabel). Obviously one of the parents must be green if we are to produce an agate, and we will start the illustration by pairing an isabel male with a green female. The chart will look like this:

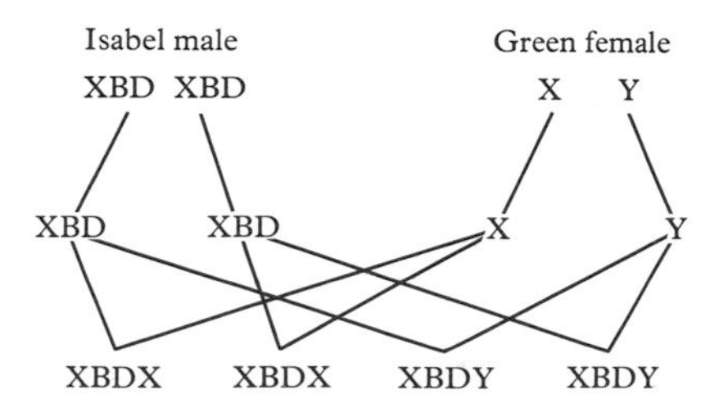

All the youngsters will be as expected, i.e. green males carrying the two factors brown and dilute, and visual isabel (brown + dilute) females. This must be so as both of the X chromosomes of the male carry both mutated genes, while the X chromosome of the female is totally unchanged and her unchanged genes are dominant over the mutated genes of the male. The female must inherit one of the X chromosomes from the male and as this is the only one she possesses both mutations must appear.

If we pair one of these young males to its sister, logically we must expect to produce normal green, normal brown, agate and isabel young. By charting this pairing we find it not to be so.

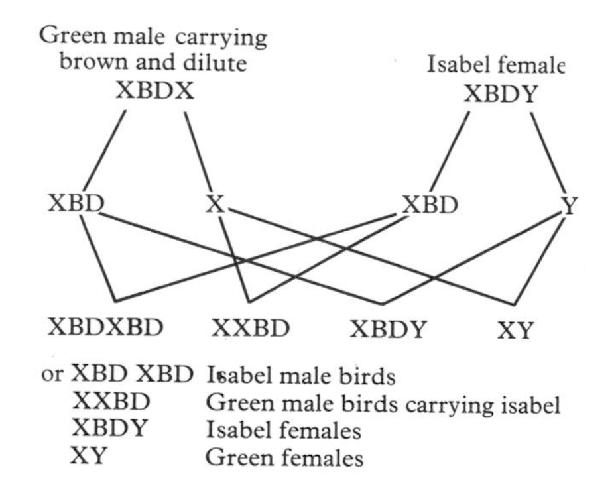

No matter how we rearrange the pairings it is not possible to produce a bird with the symbols XDXD or XD XBD or XDY, which would be an agate male or female.

To produce such a bird the dilute factor would have to be represented on both chromosomes of the male and the brown factor on no more than one. In the case of a female bird the brown factor would not be present at all.

We know from practical experience that these birds do appear from such a pairing and there must, therefore, be another answer. The explanation is known as the 'crossover'.

We examined earlier how, on the formation of the germ cells, the chromosomes appeared in pairs and then split to form the gametes. What was not mentioned was the fact that, prior to separating, they become entwined, and on separation sometimes exchange pieces. If the chromosomes are identical no effect will be noticed, but if two dissimilar chromosomes are involved the effect could be as follows:

The chromosomes could appear like this

XXXX OOOO

and when entwined like this

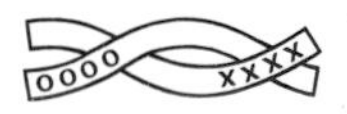

on separation they could look like

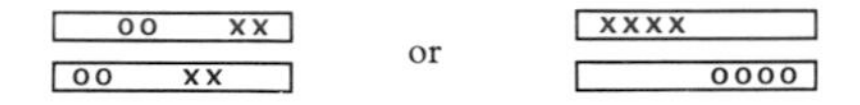

depending upon whether one or two cross-overs have occurred. One must mention that more often than not they will separate exactly as they originated.

At this juncture it would be beneficial to introduce another method of charting inheritance tables. For this we use the following letters to denote the genetical make-up of the canary: B = brown; Z = black; z = absence of black; O = oxidation factor; oa = diluting factor. All of these symbols are enclosed in a box which represents the chromosome.

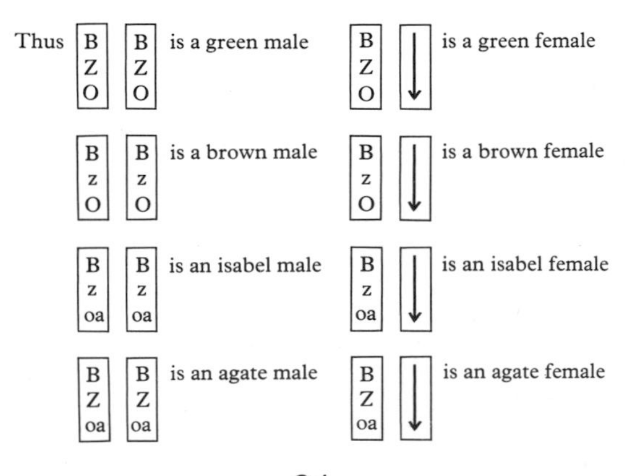

Using this new system to re-illustrate our last chart we find:

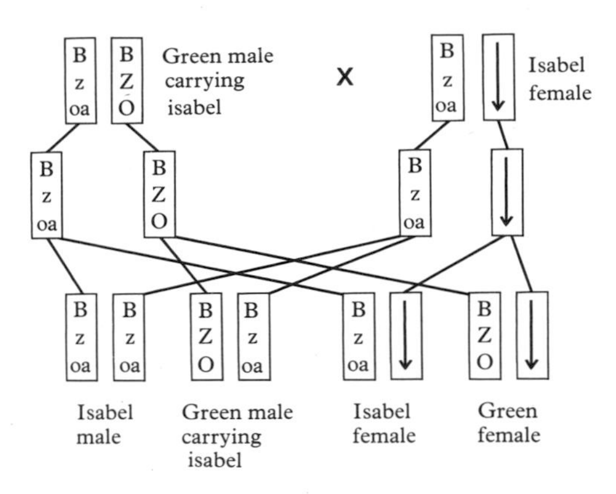

Using the facts of the crossover we can now see how the agate can appear. The green male carrying brown and isabel could on separation appear as

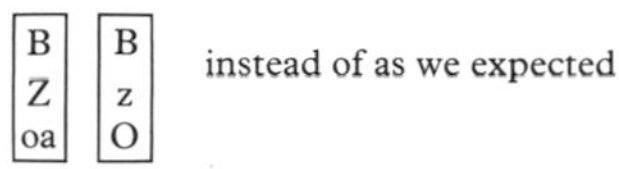

Thus when the first chromosome joins with either the X or Y chromosome of the female a male or female agate could be produced, viz.:

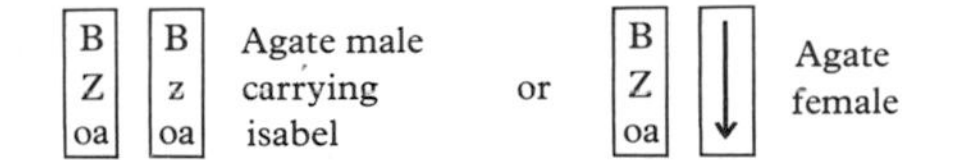

These then are the basic inheritance patterns we must follow to enable us to control our experiments and to achieve satisfaction in our hobby. Although not complete, they can be used as guide-lines to further future thought and discussion.

It must be emphasised most emphatically, before ending this section, that all expected results shown in the inheritance tables used as illustrations in this and the following chapters, must be qualified by the words 'on average'. Except, of course, where a homozygous recessive factor is paired with either its like or a homozygous dominant when the results are guaranteed. It is not uncommon for all offspring to be of one type where three types are possible. The tables simply show what is possible and over a large number of pairings what on average appears.

Hybrids

For many years experimental breeders had been producing mules and hybrids from various native species crossed either with a canary or another different variety. This was done in a rather haphazard way with the sole aim of producing something different rather than with some positive objective in mind.

When the early coloured canary breeders turned to hybridisation, their objective was very positive and precise. This was to transfer the red gene of the Black-hooded Red Siskin to the canary, thus producing a red canary.

Although other people had started experiments in this vein, notably Herr Dams and Herr Matern of Germany, it was another of their compatriots who virtually supervised and directed the early experiments, namely Dr Duncker.

It was thought that the canary had no red in its composition, and hence before a red canary could be produced, a gene that produced red pigmentation had to be obtained from some other bird and transferred to the canary. This could only happen by hybridising and the hybrid produced had to be fertile. It was known that the copper hybrid produced from the South American or Black-hooded Red Siskin (*Spinus cucullatus*) paired to a canary was in some instances fertile, this little bird having long been a favourite with South American fanciers.

The male siskin is a small bird with a jet-black hood, wings and tail, and a vermilion-red-coloured body and wing bar. The female is much duller, having no hood and only flushes of red on her breast. The red is much deeper on the rump, and the rest of her body is coloured grey.

Working on the theory that not only had the canary no capacity to produce red, but also that the siskin had no genes for yellow, it was hoped that with continued fertility, it would be possible to transfer both of the red genes of the siskin to the canary.

The first cross siskin × canary, known as the 'F1 hybrid', is a bird possessing one gene for the production of red but not yellow, with the other gene producing yellow but not red. The interaction of this red factor and the yellow factor changes the vivid red of the siskin to a copper colour.

For some reason still unexplained, when paired to a canary of any ground colour, whether it be clear, ticked, variegated or self green or brown, the siskin male usually produces copper-coloured hybrid males. These usually improve in colour after the second moult. The birds sometimes carry the hood, and wing bars of the male siskin, but almost always carry the striations of the canary.

Although fertility is common in the F1 males, only on one recorded occasion has a fertile female been found. Consequently these male birds had to be backcrossed to a canary female. Technically the application of the term 'F2', or second filial generation to the progeny of this cross is incorrect, but it is universally accepted by all breeders of coloured canaries. Strictly speaking the term 'F2' should be restricted to the offspring of an F1 × F1 pairing. A much greater variety of young were produced in this second outcross. These included copper, variegated orange and yellow, and clear or ticked orange and yellow birds. Again males were found to be fertile, although the percentage was less than in the F1 birds, but the females were still sterile. A step forward had been made, however, in that (*a*) the production of an F3 was made possible and (*b*) for the first time, clear orange birds were produced.

Using the copper and the most orange-coloured F2 males, the second backcross was made to yellow canary females. Although generally the depth of orange coloration was paler, significant inroads were made in that at last some fertile females were produced.

It was then possible to pair F3 males to F3 females using at all times the best-coloured birds available and gradually, by selective breeding to produce a strain of what were thought to be homozygous orange canaries.

When orange-coloured females were produced in this second outcross, in addition to pairing them to F3 males, they were also mated with both siskin and F1 copper hybrid male birds. In these matings, as both parents possessed hooded siskin genes the deepening of colour in the offspring although not general, was encouraging. Again fertile females did not appear until the third generation, as had been previously experienced.

Selective breeding of the better-coloured birds did not always have the desired effect, and in a number of cases the young were of an inferior colour to their parents. In other instances highly coloured youngsters were produced from inferior coloured parents. This would indicate that the production of red is not, as was at first thought, the result of a single factor, but of the combination of several factors.

In an attempt to retain more of the siskin's colour in the various hybrids, two further experiments were made. The first involved the use of a dominant white female as the outcross. It was known that the dominant white factor inhibited yellow, and it was hoped that it would not inhibit red, thus giving red and orange young. The resulting offspring turned out to be copper and ash-grey coloured, making it obvious that the dominant white mutation also inhibited red and this pairing was incapable of producing offspring more red than the copper hybrid obtained from the yellow female.

In 1929, Dr Duncker then formed the opinion that the only way to produce a red canary was via the homozygous or recessive white, colour being dominant to this factor. These theories, as we now know were incorrect, as once again the young produced were of a copper colour. This proved, however, that although the recessive white was thought to be totally devoid of yellow, the pigmentation was actually masked.

As the experiments continued two other types of coloration started to be produced. The first was a frosted bird of a pinker hue. The other, which occasionally appeared from a copper hybrid × canary mating, or, more rarely, from an orange × orange mating, was an almost white female showing colour points similar to the sexual dimorphism of the female siskin. These birds were originally rejected, but when experiments with them were undertaken it was found that they often produced redder males than had been previously obtained.

The results of these experiments are now obvious when we survey the amount of colour present in our birds over the last decade. Consequently very few experiments are now carried out using either the hooded siskin or the copper hybrid as a source of colour. We are, or at least many of us prefer to be, experimental breeders to a certain extent, and are now looking for any variation of the known pigmentation patterns that could produce something different for us. During 1975, one such breeder produced an F1 slate-grey hybrid male from a hooded siskin × silver brown canary female pairing having a most unusual wing and tail pattern which is only visible when the wing is extended. Starting at approximately 8 millimetres from the tip of the flight feathers a rectangle area centred below the shaft and covering an area of approximately 10 × 4 millimetres is present. In this rectangle either the distal or the proximal barbules seem to be absent. Speaking to some of the experimental breeders of twenty or thirty years' experience, they say that this pattern was not uncommon in some of the hybrids produced many years ago. But, as at that time the sole objective was the production of colour, its possible implications were never investigated. If the pattern can be retained and eventually transferred to the canary, it is impossible at this time to say whether or not it will be of benefit. Surely, however, we must not allow the opportunity to study this phenomenon to pass by without proper investigation. It is quite possible that secondary variations will also be present that are not obvious in the bird that exists at the moment. All eventualities must be pursued if we are to be worthy of the name of coloured canary breeders.

The illustration shows clearly the unusual marking described in the text

Lipochrome Varieties

Red Orange and Apricot

Although by far the greater number of mutations that have occurred in canaries have been on the genes responsible for the production of melanistic pigment, the popularity of clear red canaries has never wavered. In Great Britain today, these birds account for a large proportion of all coloured canaries exhibited. In many countries the original ideal of a red canary was a bird totally devoid of any melanin and even now this is the only acceptable show specimen. Other countries, however, accept birds of varying degrees of variegation although the general consensus of opinion is that an excellent clear bird will always be judged superior to a variegated specimen of equal ground colour and type. The term 'clear' is used to describe a bird devoid of melanistic pigment and the word 'ticked' is used to describe a bird with one dark mark that can be covered by an object of 1 square centimetre in area. A variegated bird is one that has more melanistic pigment than a ticked specimen, but also has some areas of lipochrome feathering visible. The term 'foul' is given to a self bird with feathers in wings or tail that are devoid of dark pigment. A self bird is one that has pigmentation in all its feathers.

There are two forms of coloration present in canaries. Firstly, the background or ground colour known as 'lipochrome colouring', which is a fat-soluble colouring material and is derived from the food which the bird eats, and secondly, the melanins, black and brown in colour, which are formed from protein produced by the bird.

Experimental breeders attempting to produce a red canary (originally visualised as a clear bird), started by crossing a Black-hooded Red Siskin male with a yellow canary female. The females from this pairing were infertile and consequently were of no use at all. The F1 male hybrids, some of which were fertile, had to be paired back to a yellow female, giving a further reduction in the number of siskin genes in the F2. In the F3 progeny some of the females proved to be fertile, and so began the complicated business of selective breeding. Birds showing as near the desired colour as possible were backcrossed and interbred in an effort to improve the lipochrome colour, i.e. red.

It is impossible for those of us who have not attempted this form of experimentation to adequately conceive the frustrations and problems involved. Regrettably too few of the early pioneers are still available for us to consult, but by talking to breeders of twenty or thirty years' experience (who at least had orange if not red females at their disposal) one can start to appreciate just how much dedication was required of the early breeders both in patience and scientific approach to pave the way for the present-day breeders who in many ways take the presence of a red canary very much for granted.

The red canary appears in two forms; these we call 'red orange' and the 'apricot'. The former is a non-frosted bird which shows a deeper colour than the latter frosted version.

At some time, possibly during its domestication, the normal frosted canary mutated to give another form of feather quality which has at various times been referred to as 'yellow', 'jonque', 'non-frosted' and 'intensive'. The respective frosted versions are called 'buff', 'mealy', 'frosted' and 'non-intensive'. The normal frosted feather is comparatively wide and shows near the body on the lower third the primary colour involved – black in the green, brown in the brown, and white in all clear versions whether genetically green or brown. This is known as the 'underfeather' or 'underflue'. The remainder of the feather in clear varieties is the dominant colour of the bird, i.e. red, yellow, pink, or white, except for the tip of the feather which is white. The lipochrome is thus partially masked by the effect of this white tip. This is known as 'frosting'. To illustrate this point better, feathers partially cover each other in the same way as tiles or slates on a roof, where at least half of each tile is covered by the one laid over it. Thus with the frosted feather laying over the red of the feather beneath it the frosting effect is achieved.

The non-frosted body feather is longer and thinner than the frosted with the lipochrome colour extending right through to the feather tip although some non-frosted birds, particularly females, still show a slight frosting. Whether this is due to the mutation not

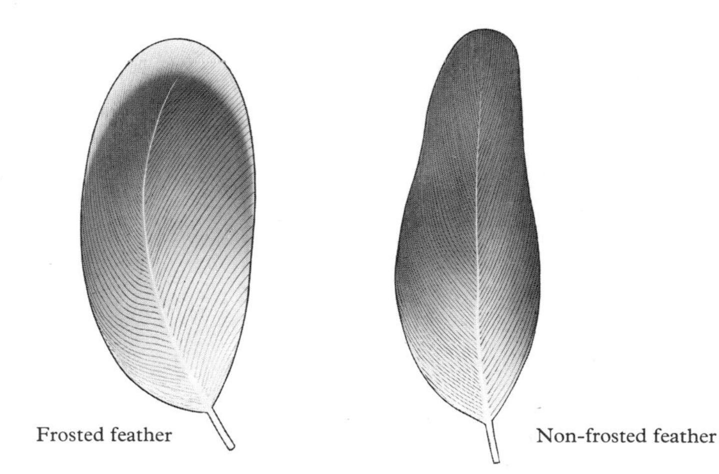

Non-frosted feather is longer and narrower. Frosted feather is wider and with white tip

being totally dominant, or is a result of secondary sex characters (which obviously only applies to females) has never, so far as is known, been ascertained. In all show specimens, the mutation should be fully expressed, giving a completely non-frosted effect in the ideal red orange bird.

The mutation that produces the non-frosted feather is a heterozygous dominant version, an example of this hereditary pattern having been examined in the section on inheritance.

From this we can deduce that a non-frosted bird paired with a frosted bird, will theoretically on average, produce 50 per cent of each in their offspring. Furthermore, a pairing of frosted to frosted can never produce a non-frosted version, but a non-frosted male paired with a non-frosted female will give theoretical average results in the Mendelian ratio, i.e. 25 per cent homozygous non-frosted, 50 per cent heterozygous non-frosted and 25 per cent homozygous frosted.

The use of non-frosted to non-frosted and frosted to frosted pairings has evoked more controversy than one would think possible, so interested is the average fancier in this aspect of breeding.

Whether the homozygous non-frosted version is viable has never been fully and conclusively established, but many varied theories have been put forward. It is commonly thought that the majority of the non-frosted offspring from a non-frosted × non-frosted pairing accentuate the effect of the mutation so that the feathers become very thin and consequently the type or shape of the bird is adversely affected. Whether this argument is substantiated by authentic records kept by experimental breeders, or has been passed down as hearsay is not important. What is clear is that we do not know the answer, and guesswork, however intelligent, is hardly the correct scientific approach. Fortunately, there are now emerging individuals and groups throughout the world who are not prepared to accept unauthenticated theories, and who are embarking on the collective studying of this and other allied mysteries. Let us hope that in the not-too-distant future we will have some accurate theses on which to base further experiments.

The use of frosted male with frosted female pairings has also produced a flurry of unsupported theories. The major one being that the amount of frosting will increase so that the ultimate result will be a bird whose frosting totally masks all form of lipochrome pigment. Until we have accurate answers to these problems any person intending to breed red canaries as a straightforward hobby and with the sole object of producing birds suitable for the showbench, would be well advised to use only the accepted non-frosted × frosted pairing. We do at least know that the majority of the progeny from this pairing will approach to some degree the accepted standards of

Top left :
Male
Black-hooded
Red Siskin

Top right :
Female
Black-hooded
Red Siskin

Right : F1
Copper Hybrid

Red Orange

Apricot

Dimorphic male

Dimorphic female

Frosted Rose

Non-Frosted Rose

Frosted Ivory Gold

Non-Frosted Ivory Gold

Clear Recessive White (right)
Clear Recessive White Satinette (left)

38

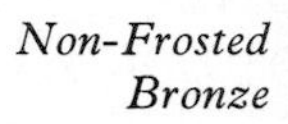

Non-Frosted Bronze

Male Dimorphic Bronze

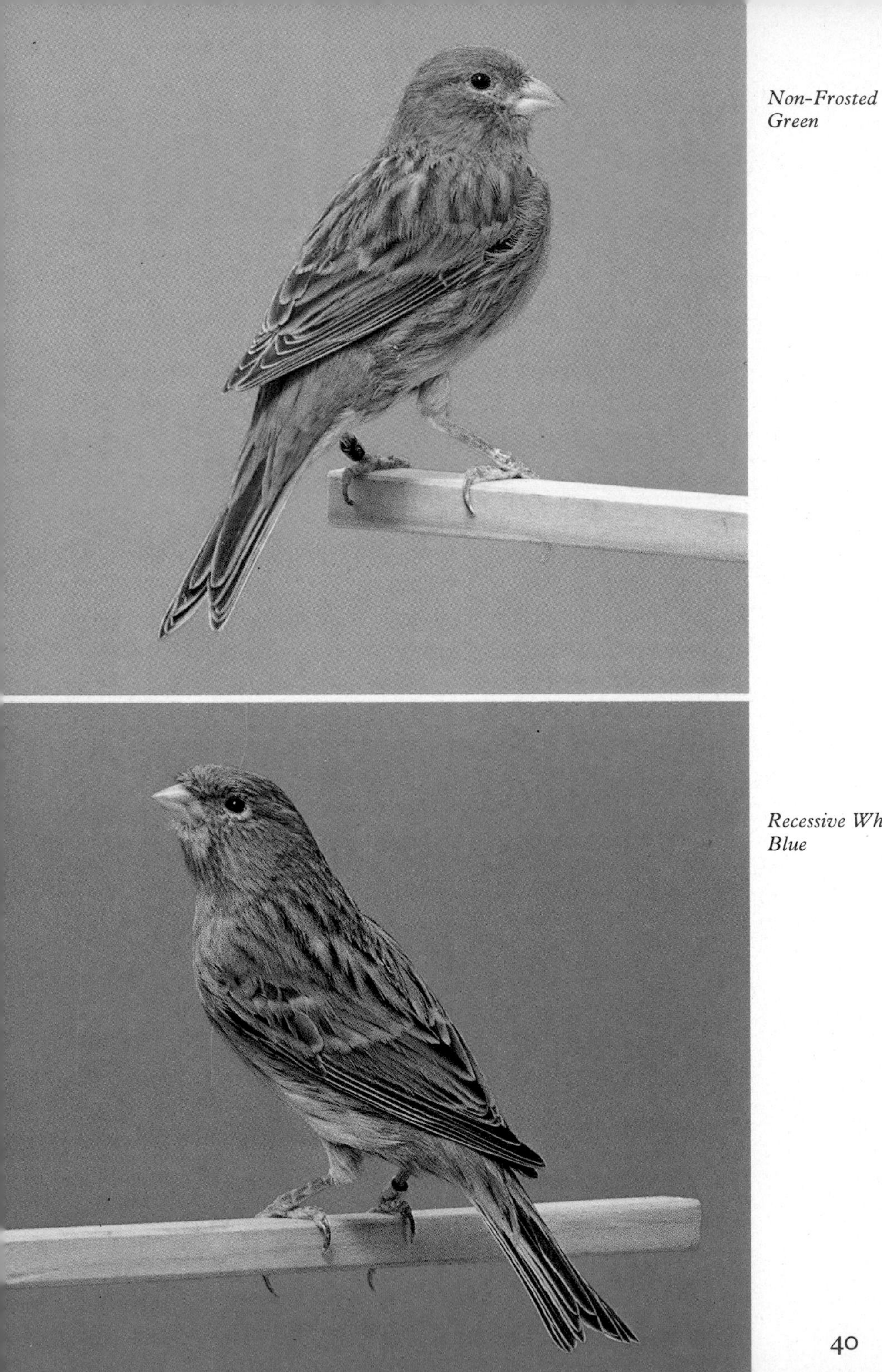

Non-Frosted Green

Recessive White Blue

Non-Frosted Rose Bronze

Silver Brown

Non-Frosted Rose Brown

Frosted Rose Brown

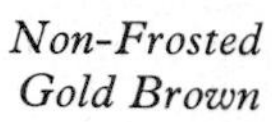

Non-Frosted Gold Brown

Pair Dimorphic Red Orange Brown

Non-Frosted Ivory Gold Agate

Silver Agate

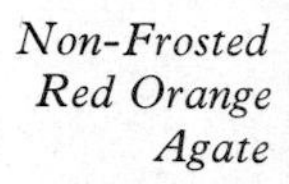
Non-Frosted Red Orange Agate

Non-Frosted Rose Agate

Female Dimorphic Red Orange Agate

Non-Frosted Red Orange Isabel

Frosted Rose Isabel

Recessive White Isabel

Frosted Red Orange Brown

Frosted Rose Agate Opal

feather quality in both red orange and apricot birds. Thus apricot × red orange or red orange × apricot matings should normally be used by the beginner. Experienced breeders with enquiring minds may attempt the red orange × red orange, or apricot × apricot pairings and may find that exceptional examples are produced. But in most instances where this has happened, the birds have been bred by breeders who have recorded over a period of years the peculiarities of each line in their strain, and the mating has been attempted with a scientific, rather than a haphazard approach.

Originally, as we have seen, a yellow canary was used to produce the red hybrids. The experimentalists basing their arguments on the theory that the Black-hooded Red Siskin carried two genes for red, and the canary carried two genes for yellow. Ultimately, it was hoped that both red genes could be transferred to produce a red canary. Their deliberations were also based on the theory that a single factor was responsible for the production of red and yellow. Over the years, the results of experiments would seem to indicate that this is not the case. If the current version of the bird is not a true homozygous red canary, then obviously red is dominant to yellow.

The normal domesticated version of the canary is a green bird, i.e. a yellow ground bird on which are deposited black and brown melanins giving the optical illusion of green. Whether or not a bird expresses melanin has nothing to do with the colour genes. The manifestation of the melanin pattern is determined by another set of genes called 'variegation factors'. Certain areas of the body of the bird are more receptive to the deposit of melanin, these are the crown of the head, the eyes and cheek, breast, flank, wings, outer tail feathers and the back. The pattern of melanin deposit is partly controlled by one of two variegation factors, and this is the major one. The second variegation factor is responsible for the depth of coloration of the melanins. The closer one gets to producing a clear bird, the further one is departing from the normal.

One ticked mark   Two ticked marks equals variegated   Typical variegated

We are able to produce clear canaries because of another mutation which suppresses the formation of melanin. To represent the normal variegation-forming factor we use the capital V and to represent its mutant allelomorph we use a small v. Thus, theoretically, VV is a self or foul bird, Vv a variegated bird and vv a clear or ticked bird. By pairing together birds with a double dose of the factor for the production of melanin, i.e. birds that have received a V factor from each of their parents, the only possible progeny must also be VV self or foul birds. The use of Vv parents paired together will produce theoretically 25 per cent VV, 50 per cent Vv and 25 per cent vv, or 25 per cent foul or self, 50 per cent variegated and 25 per cent clear or ticked young. It follows that a mating of vv parents will produce all vv or clear or ticked young.

From the variety of variegation patterns formed it is clear that neither the mutated nor the normal gene are fully dominant over the other and other factors must, therefore, be involved. No theories have been put forward to explain how these other factors work but there can be little doubt of their existence.

The amount of variegation present in a Vv bird can vary from two tick marks to an almost foul marked bird. By observing any mutation we see that if similar birds possessing mutated genes are paired together over a period of years, the offspring become progressively inferior, either weaker or smaller, or poorer coloured, or more poorly marked, or a combination of two or more of these. Why this happens is not known, but it is a common occurrence in all types of livestock. We can observe that this happens when pairing clear canary to clear canary over a period of years. Size invariably diminished, as also does colour. All breeders of red orange and apricot canaries would, therefore, be well advised to introduce either self, foul or variegated green birds into the clear stock in alternate years. The use of brown stock is not advised for correcting these faults.

As can be seen from the tables below a number of clear young will normally be obtained from two of the three featured pairings, and these will normally be superior to clear young produced from a clear × clear pairing.

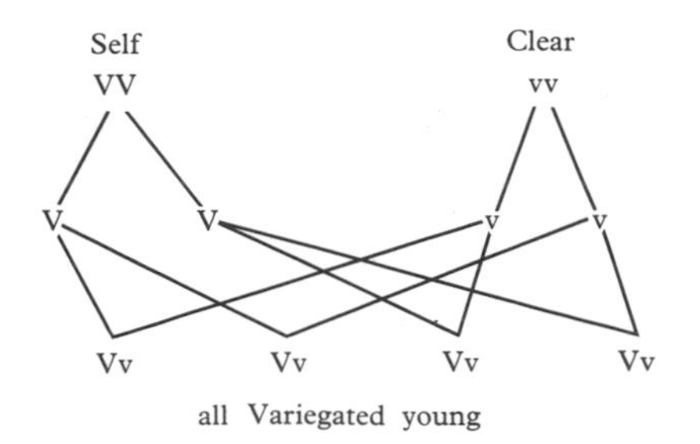

all Variegated young

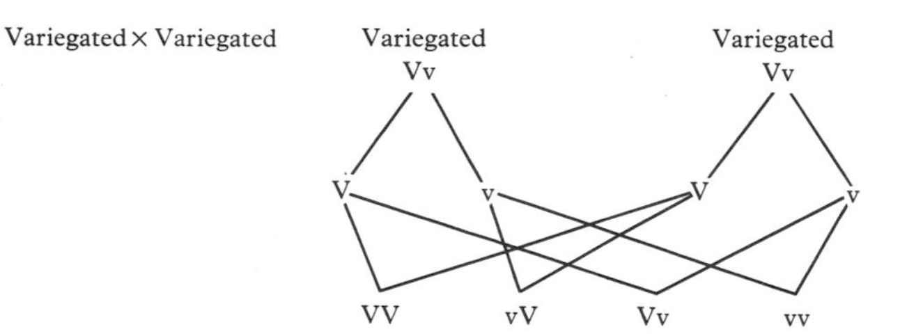

25 per cent Foul or Self (VV) 50 per cent Variegated (vV Vv)
25 per cent Clear or Ticked (vv)

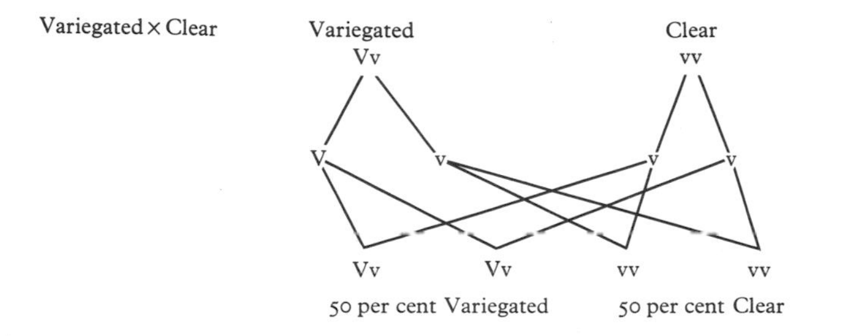

All percentages are those which can theoretically be expected.

At varying times both brown and dimorphic birds have been used to help in producing higher density red birds. The results of the former are at best inconclusive and the latter haphazard. When the dimorphic female first made her appearance she was used as a partner for hybrid males. Some birds produced from this pairing proved to be of superior colour, but no published results are available concerning the quality of the remaining majority of the progeny. In the 1950s and 1960s dimorphic females were universally used in the hope of discovering some sort of magic solution for the production of red lipochrome. However, when the rampant factor responsible for the wider feathered sex dimorphic became effective throughout the stock it resulted in frosted red orange and very heavily frosted apricot birds. Breeders then quickly came to their senses and, realising that they were losing considerably more ground than they were gaining, promptly dropped the idea. It took several years for the genes responsible for the dimorphic pattern to disappear even though most breeders were discarding all birds showing the undesirable pattern. Today it is rare for the factor to appear in established red orange/apricot strains.

The breeders of red canaries up to the mid 1960s used no form of artificial colouring agent to assist in the production of visual red, and, even on its introduction, were still of the opinion that the red canary had yet to be bred.

In retrospect we must ask if these comments were correct. The ground colour of the canary is dictated by genes responsible for giving the signal for the production of colour. This coloration is derived from food that the bird has eaten. The normal canary gets its yellow colour from carotenoids contained in its food, the yellow apparently being derived from a yellow pigment called 'xanthophyll', while the red results from a carotenoid called 'carotene'. Both of these carotenoids are present in a wide variety of foods. Whether the gene for red is capable of using both is doubtful, but it is known that the gene for yellow does not use carotene as a source for colour. If a bird is deprived of a supply of food containing carotenoids it will go white regardless of its ground colour.

It is known that the siskin male and the copper hybrid will lose colour if not given a diet containing carotene. This suggests that in part, at least, the red of the siskin is dependent as much on its food-supply as on its genetical make-up. Within most of the countries where the breeding of the red canary is prevalent, few naturally grown foods contain high quantities of carotene. Of all vegetable products analysed only six showed a carotene content of over 100 parts in 1,000,000. Of these six only three were readily accepted by birds. These are dried grass which has a carotene content of between 160 and 250 parts per 1,000,000, carrot root with 120 parts per 1,000,000 and grasses with 117 parts per 1,000,000. Unfortunately the analyst did not record what varieties and in what state the examples were analysed. Is it any wonder then that a visually red canary was not produced even though comparatively huge quantities of these foods, particularly carrot, were fed to the birds?

Today, the use of a manufactured colouring agent is acceptable in most countries and, as these agents have a carotene content of approximately 10 per cent, the task of producing a visually red bird is eased considerably. When these were first introduced the general feeling among fanciers was that such artificial methods of feeding would prevent accurate assessments to be made of the birds' ability to produce red. Tests on yellow and white ground birds, however, have proved conclusively that unless a canary possesses the gene for the production of red it will not moult out red whatever the quantity of colouring agent offered.

Certain birds, when fed on these colouring agents during the moult will subsequently appear a dull red, purple or brown, rather than the bright red colour so hopefully sought. It has commonly been thought that this was due to the overfeeding of the colouring agent, but the writer is of the opinion this is incorrect. Over the last two years experiments in colour feeding have been conducted which, while not conclusive, do show certain patterns which suggest that when a bird moults out overcoloured, it is not due to its diet but to its genetical make-up. The factor believed to be responsible is the optical blue reduction of brown factor. Not one, but many genes, are responsible for the genetical make-up of the black and brown factors, and on separation to form the germ cells a set number of these genes pass in a random manner into the gametes for forward transmission. When a high percentage are brown genes, the lipochrome will appear dull, and when colour fed will give this purplish-brown effect. When a high percentage are black genes the lipochrome will appear luminous.

Certainly at this stage no theory can be put forward as a thesis, but if experiments continue to follow similar patterns for a number of years then perhaps these theories can be either proved or more readily accepted.

Another point that must be mentioned is that no bird can show colour deeper than its assimilation peak. Only the actual inheritance of genes controlling colour can affect the amount of colour that can be maximised in a canary, and a poorly bred bird can never achieve the colour of a better bred bird.

So complex are the factors governing colour that at no time can we determine the depth of colour which the progeny will possess by studying one pair of birds. It is, therefore, vital that accurate and extensive records are kept of the type and colour patterns that emerge in a strain. Some pairs may produce top-quality apricots but poor red orange specimens. Others may produce not only poor apricots showing too much frosting, rather than the even distribution of frosting required, but also red orange birds showing no frosting whatever. Only by recording all these details can we start to scientifically breed for a known and desirable end result.

Before leaving the subject of coloration, one must make mention of two other points. Firstly, the effects of environment. Although again no factual theses have been presented it has long been thought that birds moulted out in bright sunlight are partially bleached and do not display as deep a colour as those moulted out in an environment where bright sunlight is excluded. In the absence of contradiction this is one aspect that may well be taken into consideration. Secondly, it must be remembered that the bulk of the diet of a canary consists of various seeds, and it was here, certainly prior to the introduction of manufactured colouring agents that the breeder of red canaries was able, to some extent, to exclude from the diet any yellow-producing carotenoids. For example, rape, thistle and hemp contain comparatively large percentages of lutein, a yellow-producing carotenoid, while oats contain a very small amount, and maw and niger seeds contain none. These three seeds were then the only ones offered, and there can be no doubt that they did help the birds to show improved colour. With the introduction of the manufactured agents the same necessity no longer existed, but experiments still need to be carried out to give a factual answer.

Until we have positive answers to these experiments, breeders who are anxious to succeed with their birds on the showbench may care to remain true to the traditional seed diet of niger and groats or pinhead oatmeal while the bird is moulting.

Much still needs to be researched before we can attain consistency in our aims, let us hope that the spirit of adventure and the thirst for knowledge will not be exhausted before we know all the answers.

The Dimorphics

Classic or Old Type

We have already referred very briefly to the appearance of the dimorphic canary. Occasionally a very pale type of female appeared from early red canary pairings, which at the outset was discarded. This must be one of the few occasions when we can accuse the early breeders of shortsightedness. Their sole aim, however, was the production of red, not the establishment of a new variety, and it was, therefore, felt at the time that females so low in colour would be of no use for breeding. Eventually the German breeders did experiment with the birds, and they established a strain of red canaries that were called 'carmines', so strong was the red colouring compared with that so far achieved. Nowadays, as we do not have to search so intently for red pigment, we can instead examine the dimorphic variety separately. So great are the disadvantages of using dimorphics, as explained in the section on red orange and apricot varieties, that it is considered most detrimental to introduce this species into our clear stock.

The Black-hooded Red Siskin male, from which red canaries are derived, is a highly coloured bird, while the female is mostly grey with some colour points. These colour points are located on the rump, wing butts, upper breast and round the eyes, the colour being a paler version of the red of the male bird. This difference is known as 'dimorphism' and is not attributable to the genes for colour, but to separate characteristics. There are certain characteristics which are associated with sex because the genes responsible for them are situated on the sex chromosome. Other characteristics, which are visible in individuals of one sex only, are known as 'sex limited'. These factors do not function on a sex-linked basis because the genes responsible for their appearance are not situated on the sex chromosome but are on the somatic chromosomes. Nevertheless the normal hereditary patterns do not emerge, and the characters involved appear in one sex but not in the other. It has been suggested that the reason for this is that either testicular or ovarian hormones play a part in restricting these characteristics to one sex only. The difference in coloration of the male and female siskin is one instance of this.

It is evident that in the case of the dimorphism of the Black-hooded Red Siskin the factor is partially sexual, which leads us to believe that the male Black-hooded Red Siskin carries the factor involved. On pairing a siskin male with any type of canary female the female F1 hybrids produced almost always appear very similar to the female siskin, i.e. grey with colour points.

Whether the gene that causes this effect in the Black-hooded Red Siskin has been transmitted to the dimorphic female canary or whether the effect is caused by one or more different factors in the canary operating either independently or in conjunction with the genes from the siskin, has not yet been established. What is obvious is that it cannot be the effect of a single gene because a dimorphic specimen can suddenly appear in a strain of canaries that has produced no dimorphics for years.

Originally the dimorphic pattern emerged in a female canary. This bird was similar in appearance to its red orange and apricot nest mates prior to its first moult, but on moulting lost rather than gained colour. On completion of the moult the bird resembled a white canary with orange colour points located in similar positions to the colour points of the female siskin. Breeding traits of these original birds are difficult to ascertain because, as mentioned before, the production of red was the only objective and few, if any, breeders have recorded details of their breeding results.

We do know, however, that it was thought that the dimorphic characteristic was limited to the female and consequently no serious thought was given to the male birds used in matings.

A few breeders, however, did experiment with the dimorphic, and produced a theory which at that time was so startling that leading breeders of the day became increasingly sceptical. The theory was that a male version of the dimorphic female existed.

Up to that time it had been considered impossible for a male dimorphic to exist, but the evidence put forward seemed partially conclusive. The bird that appeared resembled an apricot male except that it carried more frosting, particularly round the neck and along the back, it also had deeper colour points. What was especially noticeable was that the area of white feather round the vent was extended backward and forward. Whether such birds were true dimorphic males or merely poor apricot specimens partially influenced by the gene(s) responsible for the production of the dimorphic females is still a matter of speculation. What is known is that when paired to dimorphic females these birds produced female progeny that approached the ideal for which breeders were striving.

Although presumably a separate factor, when successive pairings are carried out between unsuitable partners the dimorphic pattern starts to diminish allowing the lipochrome to be more evident and eventually the birds begin to look more like poor-quality apricots. Thus, to breed good-quality specimens of the original dimorphics (known as 'classic dimorphics') an apricot male, showing as heavy frosting as possible, should be paired to a female as near to white as can be acquired. The male produced from this pairing which shows the highest amount of frosting (particular regard being taken of the white area in the region of the vent) should be paired back to its mother. Future generations should be similarly paired, using vigorous selection, until a true breeding strain of dimorphics is achieved.

The feather of the dimorphic is distinctive in that it is much longer and broader than that of the normal frosted canary, and it appears in two forms. The first is a true white feather. This is usually situated round the vent and flank areas. The second form appears in two sub-forms. Firstly, as an extended frosted feather, i.e. white on the lower third to half and on the tip, with lipochrome colouring between. The frosting on the tip is greatly extended when compared with the normal frosted feather, so that the colour is limited to a small area in the centre. It is the overlapping of these feathers that give a white effect throughout the body. Secondly, the feathers in the area of the

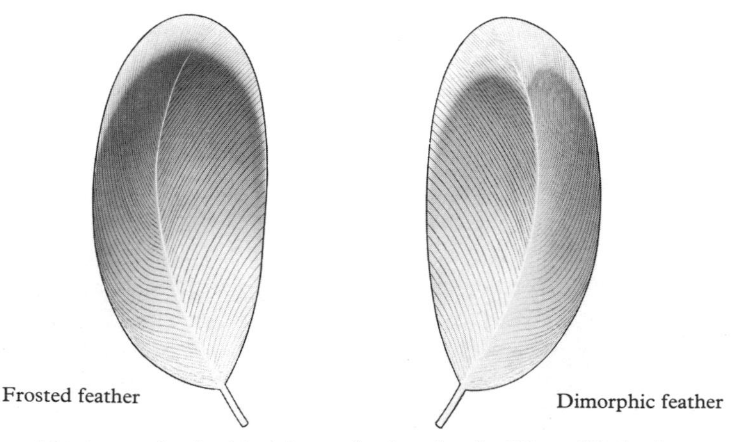

Note extra depth of frosting on feather tip of a Dimorphic feather

colour points, which although frosted at the tip, have a greatly reduced amount of frosting giving a solid colour effect. With this in mind it is obvious that the introduction of the mutated non-frosted feather, via the red orange, would be detrimental.

It is ironical that the dimorphic factor although originating from the siskin, which is a self bird, has really only been researched, with any success, in the clear series. With the classic self birds (the green, brown, agate and isabel) a great deal of the beauty is derived from the lipochrome colouring. With the introduction of the various mutations however, causing delicate alterations to the melanins which can be partially masked by a rich ground colour, there is great scope for experimenting to establish the dimorphic factor in the self varieties.

New Type

We will now proceed to study the new type dimorphics and again most experiments have been carried out on clear varieties.

During the late 1960s there appeared via an Italian breeder an apparently new form of dimorphic canary. The females in general appeared perfect, being snow white with the lipochrome coloured points both distinct and limited to the designated areas. This was unlike even the good classic specimen in which the lipochrome always tended to extend into the wings and round the head, instead of being restricted to a red extension of the eye and an area across the chest. So perfect was the top-quality specimen in fact that the colour appeared to have been painted rather than inbred. Even more exciting was the male form which had a background colour of light pink, white areas on the vent, flank and inner chest, and bright red colour points. One of the more striking effects was the mask which resembled that of a Goldfinch (*Carduelis carduelis*). So different were these birds when compared with the classic dimorphic males that, on seeing an example for the first time, one could be forgiven for thinking it to be a poor-quality classic dimorphic female.

The genetical behaviour is also different from that of the classic, which makes the new type an even more welcome addition to the range of coloured canaries.

Although these birds have only been in Britain for three or four years, certain breeding patterns have already started to emerge. By pairing a new type male to a new type female all new type young are produced. It must be made clear that not all new type dimorphics are perfect examples. Breeding skills are just as important with this variety as with all others. A new type male paired to a classic female, gives extremely white females which are assumed to be new type, but could be either well-formed classics or some form of intermediate between the two types. The male progeny appear similar to very poor apricots and when paired to a known new type female give a mixed variety of young of both sexes. This would seem to indicate that the factor is to a certain degree governed by sex linkage, and could be a partially dominant homozygous recessive factor, although it is as yet too early to be certain.

Two other points have emerged from the writer's limited experiments which could be simply related to his own strain and of little consequence when studying the variety generally. It will be interesting to hear if other breeders have had similar experiences. The first point is that the new type dimorphics seem to be less robust than other varieties. They are always the first to show any form of stress, and are invariably the last birds to come into breeding condition. Why this should be so is not apparent, but if widespread obviously invites further experiments in an attempt to isolate the cause and correct it. Perhaps these birds need some form of dietary additive as do the recessive whites. The second point is the problem of the appearance of lumps. The cause of lumps in canaries has been discussed over many years, and several theories have been put forward to explain it. Although knowing the effect, which is the formation of a feather cyst at the place where the feather fails to break the skin forming a large unsightly lump, it would seem that both the cause and treatment of the fault is unknown. It has long been thought that the reason for the appearance of lumps is the continued pairing of frosted to frosted, causing the feather of the young to become long, wide and soft and, therefore, unable to break the skin. This feather make-up certainly conforms with that of the dimorphic, but unfortunately does not explain how lumps can be found on non-frosted birds. Lumps are normally found in only three varieties of canary: the Norwich, the Yorkshire and the Gloster, all of which have been inter-mated to produce the current versions. The new type dimorphic canary has some Gloster canary ancestry, and it is possible that the lump effect is in fact a genetical mutation which is passed on by a hereditary pattern. So far as is known there are no recorded instances of lumps appearing in classic dimorphics.

By pairing new type dimorphic to new type dimorphic over a period of years the progeny certainly appear to become whiter. Whether this is due to the feather becoming wider, thereby obscuring the lipochrome or whether the lipochrome gradually diminishes is not known. The writer, however, did carry out this pairing for three years, produced by far the whitest chicks yet seen but also was faced with the problem of lumps. One male with lumps was paired with a female showing more extensive red areas than a characteristic example, and the resultant offspring showed no trace of lumps. Time will tell if this is the answer to the problem, but it suggests that after two years of pairing new type to new type, either a new type dimorphic male should be outcrossed to a classic dimorphic female, or a classic male to a new type dimorphic female. With continual new type dimorphic × new type dimorphic pairings the colour points tend to become less evident owing to the overlay of the wider white feathers. The resultant loss in feather width in the progeny of the two outcrosses suggested as a possible preventive for lumps also allows the colour points to be better expressed, which is a secondary benefit, but one that should be kept in mind.

The origin of the new type dimorphic is vague, but it has been confirmed that the Gloster canary is involved in its creation, and this is fairly obvious from the shape of the head of these birds. By studying Gloster canaries on the showbench it is not uncommon to see buff specimens that carry a positive dimorphic pattern. It is thought that the Italian breeder who introduced the new type dimorphic also noticed this trait and proceeded initially to selectively breed Glosters showing a reduction of lipochrome. In all probability what actually occurred was that as the feathers became wider the overlapping of the white frosting masked the lipochrome. It was then reasonably easy to introduce the red gene by way of a classic dimorphic female, and the new type dimorphic was created. It must be emphasised that this is largely guesswork and should not be taken as conclusive fact.

Dimorphics, either classic or new type, need no special attention when colour feeding, the same strength and quantities being offered as to the red orange and apricot birds. The distribution of colour is determined not by the amount of carotene consumed but by the genes responsible for colour distribution.

The ideal female showbird, whether clear or self, should only display lipochrome pigment (*a*) as an 'eyebrow' with the colour running neither from eye to eye, nor down to the cheeks; (*b*) as a small distinct area on the wing butts that does not extend to the flights; and (*c*) with the colour on the rump but limited to that area and not allowed to extend to the back or underbody. Ideally the chest should be totally devoid of lipochrome pigment, but it is only on very rare occasions that this is achieved without a resultant loss in one or more of the colour points. In those countries where it is acceptable for male birds to be exhibited, the colour points should be as for the female except for the face where the goldfinch-type mask, distinct and bright, should be present. Elsewhere, the bird should be as white as possible, but it must be remembered that this will never be so distinctive as in the female.

The Ivory

We have examined the various factors involved in the search for red to the exclusion of yellow and will later study the appearance of white plumage to the exclusion of all lipochrome colours. It is now opportune to consider the only factor that modifies all of these lipochrome factors, namely, the ivory.

This is a sex-linked recessive factor which is subject to the normal inheritance pattern associated with the sex chromosomes. It follows, therefore, that a male in which one of the two lipochrome-producing genes has mutated will have the same phenotype as a homozygous normal but will be capable of producing among its offspring both visually mutated females and also normal males of which half on average will carry the factor. A female with just one sex chromosome must be either a normal or a mutated specimen.

Exactly when the mutation first occurred is difficult to determine, but was about 1950. A Dutchman, the late P. J. Helder, was the breeder in whose stock it appeared. From a normal red orange × apricot pairing a pale coloured chick was produced. On moulting, this bird, which proved to be a female, turned a pink colour, rather than the red that was expected.

This version, i.e. the modified red we refer to as 'rose' (formerly rose pastel) and its yellow and white counterparts are known as 'gold ivory' and silver ivory'. When early specimens of the bird were imported into Britain, there also appeared a bird described as being the 'colour of old piano keys', this was called the 'ivory pastel'. A lot of controversy has surrounded this bird, the major problem being to determine exactly what genetical make-up it had. We know from experiments carried out since that it is not an ivory red, yellow or white, as the bird in no way resembles the description given, and we must turn to records of the day to give us a clue. The ivory pastel was referred to as a bird with a lower red gene content than the rose, and we must also remember that at the time of its appearance, carotene was only fed in its natural state, e.g. carrot roots. If a red orange is paired to a frosted yellow, the progeny usually are a pale orange colour, when not colour fed. If the ivory factor is introduced the orange colour turns into a dirty yellowish orange which is as near as we can get to the 'old piano key' colour described. Except for obscure experimental work this version can be of no benefit to colour breeders and in consequence beginners would be well advised to concentrate on the pure ground colours.

As has been stated, red becomes modified to pink, and the yellow ground version appears as a pale lemon-coloured bird. A white, whether modified or not, still appears white, but here the factor has other effects. One of the most obvious things with a dominant white is the yellow lipochrome colouring present in the flights. When the ivory factor is added these depositions of lipochrome disappear, but if examined closely a very pale suffusion of lipochrome can be seen distributed throughout the whole of the body plumage. This is normally very difficult to see unless the bird is then colour fed. In the majority of normal dominant whites, regardless of the amount of carotene fed, the bird will remain white. Almost all silver ivories when fed carotene will turn pink in certain areas of the body.

With all birds the depth of colouring is dependent upon the genetical make-up, a bird with a low lipochrome-producing gene content will not achieve the depth of coloration of a bird with a high lipochrome-producing gene content. This is equally true with the ivories. A bird can never produce a greater depth of colour than its assimilation peak, regardless of the amount of carotenoids fed to it. The red version of the mutation is by far the most popular with breeders, and, when first introduced, the ideal was described as being the colour of a wild rose. The ambiguity of this description caused British breeders to attempt to define a shade of colour for which to aim, but this met with innumerable problems. Paint colour charts were studied at length to find the ideal but, in every instance, anything that approached the popular conception of the required colour was found to have blue in its composition. As this colour does not appear in canaries, the situation was accepted to be impossible and a more sensible, although equally controversial compromise, was reached. With the red canary, great emphasis is placed on the attainment of the ultimate in depth of colour, and this aim has been adopted for the rose. Our standards now give preference to birds showing as deep a shade of pink as possible. It must be emphasised that this will only be achieved by breeding birds with the correct genetical make-up and not by increasing the amount of carotene fed.

One interesting side-effect of the mutation that is worth mentioning is that an evenness of colour is automatically achieved, unlike the normal red, where the shoulder butts and rump will usually appear slightly deeper in colour.

All the normal rules for breeding clears apply to the ivory mutation, i.e. ground colours should not be mixed, frosted birds should be paired to non-frosted and, as with all mutations, the ivory should be paired back to a normal at least in alternate years. It is just as difficult to retain depth of colour in rose birds as it is with red orange and apricot specimens and thought must be given not only backcrossing but also to introducing self or variegated stock. This will, in addition to assisting with colour, also help to prevent loss of shape and size. We have seen that the red orange exhibits a more solid colour than the apricot because of its mutated thinner non-frosted feather. The non-frosted rose equally possesses a more solid colour than its frosted counterpart, and the normal rules for total absence of frosting in the non-frosted version and an over-all finely frosted effect in the frosted version apply.

Although we have only described the rose in a frosted and non-frosted clear form, the mutant can and does exist both with the dimorphic pattern and in the self series. In the latter form it is particularly sought as a means of producing really beautiful shades of colour.

The dimorphic rose is not so outstanding as its red counterpart simply because there is less contrast between the two colours, namely pink and white. Notwithstanding this, the dimorphic rose is a most beautiful bird of very delicate colour and while, when in direct competition with the red variety on the showbench it is invariably overshadowed, it is nevertheless a welcome addition to any experimental breeder's birdroom.

As the mutation occurred in clear stock, the transfer of the factor to self birds was achieved by outcrossing initially to a self, whereby variegated young were obtained. With the production and suppression of melanin factors playing their part, eventually a self version was bred, whereupon its introduction to the various range of other mutations was carried out. In several of the mutated self varieties the beauty of the various factors can best be appreciated when they are superimposed on a pale ground colour, and it is here that the rose, and to a lesser extent the gold and silver ivories, come to the fore.

The basic inheritance table for the mutation is as follows:

1 Ivory male × normal female produces normal males carrying ivory and ivory females;
2 Normal male carrying ivory × ivory female produces normal males carrying ivory, ivory males and females and normal females;
3 Normal male × ivory female produces normal males carrying ivory and normal females;
4 Normal male carrying ivory × normal female produces normal males carrying ivory, homozygous normal males, normal females and ivory females;
5 Ivory male × ivory female produces all ivory young.

Note: Ivory denotes either rose, ivory gold, ivory silver. Normal denotes either red orange or apricot, gold or silver.

Dominant and Recessive Whites

Prior even to the list of known varieties published in France in 1709, when the brown was first recorded, there were reports of white canaries existing in Germany. Of their genetical make-up we know nothing, but today there are two forms of white canary in our hands, the inheritance patterns of which we do have knowledge.

The first of these is a homozygous recessive version known as the 'recessive white'. The first recorded example of this mutation was a ticked bird bred from a pair of frosted yellow birds, by a Miss Lee of New Zealand in 1908. The bird changed hands and various strains were produced most of which seem to have become extinct within ten years. Someone, somewhere, must have persevered however, as the mutation is not only alive and well but gaining new supporters yearly, both in clear and self forms.

The action of the recessive white factor when the mutation is present in double dose is to totally prevent the expression of any form of lipochrome pigment. The skin colour of the bird also changes from the normal pink to a lilac colour.

As the mutation is a homozygous recessive, its inheritance pattern is not subject to sex linkage. Both genes of the pair producing lipochrome need to have mutated before the characteristic can express itself. Possible matings and theoretical expectations are as follows:

Pair 1 Recessive white male × recessive white female produces all recessive white young.

Pair 2 Recessive white male × normal female produces all normal young carrying recessive white.

Pair 3 Recessive white male × normal female carrying recessive white, produces 50 per cent recessive whites, 50 per cent normals carrying recessive white.

Pair 4 Normal male × normal female carrying recessive white produces all visual normal young, 50 per cent of which carry recessive white.

The males and females in the above can, of course, be reversed to give identical results. The word 'normal' is used to identify any lipochrome colour.

In its early history it is thought that the recessive white was bred only as a clear, but eventually it was transferred to produce recessive white blue, brown, agate and isabel specimens. As the factor only affects the lipochrome this would seem to be a comparatively easy task. Initially a clear recessive white male was paired with a self green female. The green colour in this bird is caused by a superimposition of black and brown melanins on a yellow background. These same melanins transferred to a white ground bird (now referred to as 'silver', e.g. silver brown, silver isabel) give the appearance of a slate-grey bird which is called a 'blue'. The pairing of the clear recessive white with the self green produced variegated green birds all carrying the factor for recessive white. The amount of variegation, its pattern and depth, was determined by the variegation and pattern factors described more fully in the section on red orange and apricot. These birds were then heterozygous yellow specimens possessing one gene for the production of lipochrome and one mutated gene. The former being dominant over its mutated partner, the ground colour of the bird was not affected. The birds also carried one gene for the production of melanin and another mutated gene for the suppression of melanin. When attempting to transfer one recessive mutation from one species to another the task is relatively simple. When more than one is involved, however, it is clear that the problem is greatly increased, and needs much patience and tolerance if the breeder is to achieve his aim. To give some indication of what is involved, a chart is reproduced below. The capital P is used to represent the dominant yellow gene and the small p the mutated recessive white gene. The capital V is used to represent the factor producing melanin and the small v the mutated version supressing melanin.

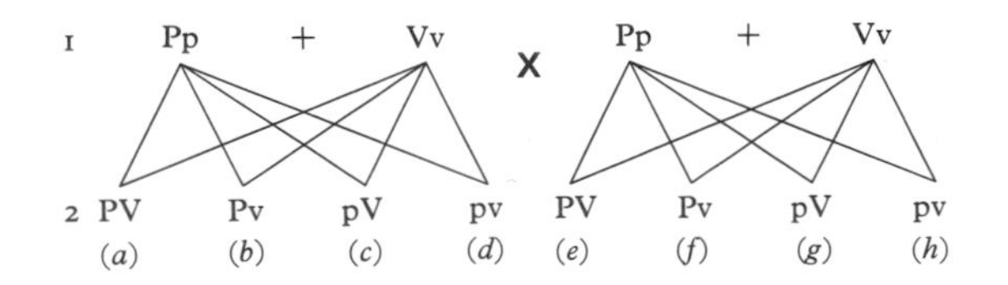

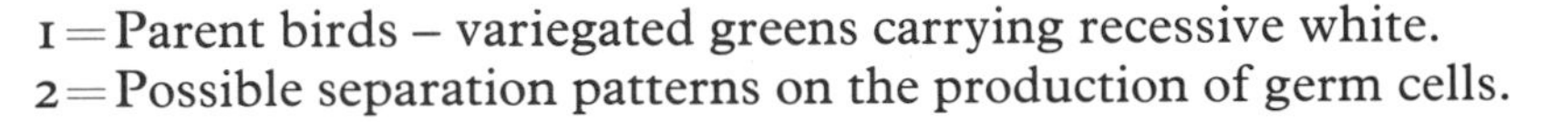
I = Parent birds – variegated greens carrying recessive white.
2 = Possible separation patterns on the production of germ cells.

When the male and female gametes join to form the zygote the following patterns could emerge.

(*a*) could join with (*e*) to give PVPV (or PPVV) which is a self homozygous green.
(*f*) to give PVPv (PPVv) which is a variegated homozygous yellow.
(*g*) to give PVpV (PpVV) which is a self green carrying recessive white.
(*h*) to give PVpv (PpVv) which is a variegated yellow carrying recessive white.
(*b*) could join with (*e*) to give PvPV (PPVv) which is the same as (*a*)+(*f*).

(*f*) to give PvPv (PPvv) which is a homozygous clear yellow.
(*g*) to give PvpV (PpVv) which is the same as (*a*)+(*h*).
(*h*) to give Pvpv (Ppvv) which is a clear yellow carrying recessive white.

(*c*) could join with (*e*) to give pVPV (PpVV) which is the same as (*a*)+(*g*).
(*f*) to give pVPv (PpVv) which is the same as (*a*)+(*h*).
(*g*) to give pVpV (ppVV) which is a recessive white self blue.
(*h*) to give pVpv (ppVv) which is a variegated recessive white.

(*d*) could join with (*e*) to give pvPV (PpVv) which is the same as (*a*)+(*h*).
(*f*) to give pvPv (Ppvv) which is also the same as (*b*)+(*h*).
(*g*) to give pvpV (ppVv) which is the same as (*c*)+(*h*).
(*h*) to give pvpv (ppvv) which is a clear recessive white.

There is, therefore, a theoretical one in sixteen expectation of producing recessive white blues. It must be emphasised that this is a theoretical expectation. It is possible, but unlikely, that all of the young will have the genetical make-up pVpV, i.e. recessive white blues or, alternatively, any of the other combinations, the actual offspring produced being totally determined by the way in which the chromosomes separate into the germ cells and re-join on formation of the zygote.

Once we have the mutation established in the self series it is then comparatively easy to reproduce it. The one unsatisfactory feature of the blue canary is that on a white background the phaeomelanin brown which is present on the tip of the melanistic feather is more pronounced giving a smoky rather than a luminous grey effect. To eliminate this fault is very difficult and may not even be possible particularly in females because they show more brown than male birds. The only means at our disposal is by selective breeding to try to introduce the optical blue factor. With the availability of the recessive white blue it became possible for breeders to introduce the sex-linked agate, isabel and brown factors into their birds to give a further range of colours with which to experiment.

As was mentioned earlier, the factor first came into prominence about 1910 which was prior to the start of the search for red. The only ground colour available before that time was yellow, and all that could be known was that the factor was dominant to yellow when in double dose. Later, when the search for red was well established, breeders used the recessive white to pair to the Black-hooded Red Siskin hoping that the factor would not be dominant to red. Unfortunately, as we now know, their hopes were dashed and it was proved beyond doubt that the factor is dominant over any form of lipochrome colouring.

Most coloured canaries carry a high percentage of points for colour when being judged at exhibitions. With no colour possible in a clear bird, a nominal number of points are automatically awarded for lipochrome, with the balance being awarded for type, feather quality, general condition, etc. In the self series it is, of course, possible to award points for the presence, dilution or absence of melanin depending upon the variety.

One point not mentioned with this mutation is its inability to convert vitamin A from its food intake. The failure of early breeders to recognise this could be one of the causes of their problems. It is easy for us to overcome the problem now that we know of its existence, using either a vitamin concentrate in the seed or water, or the more historical method of letting the birds have constant access to beef suet.

In their clear form these birds are not a suitable subject for beginners, being extremely fragile until the first moult has been completed. At all times the breeder is advised to use a normal bird carrying the factor as one of a pair, as the offspring of a recessive white × recessive white pairing seem to be even more delicate. With selfs the problem is less formidable with the chicks from either pairing being quite strong and vigorous. To eliminate possible problems and disappointments, however, it is still advisable to use normal birds carrying recessive white as mates for the progeny from a recessive white × recessive white pairing.

The dominant white appeared on the showbench in the early 1920s in Germany. Whether it was a descendant of the whites recorded in Germany in the 1600s or was a later mutation is unknown. As its name implies, this version of the white canary is a dominant heterozygous and unlike its recessive counterpart is dominant to any lipochrome colouring, consequently a bird is either dominant white or not. It is not possible to have a bird carrying the factor. When paired to a yellow female, a dominant white male either unites its unmutated yellow-producing gene on its colour-suppressing gene with one of the female's yellow-producing genes. On average 50 per cent of the offspring will appear white, the other 50 per cent yellow.

By mating two dominant whites together twenty-five per cent of the progeny will on average be homozygous dominant whites, 50 per cent heterozygous dominant whites and 25 per cent yellows.

This last-mentioned pairing is not recommended, however, as the mutant when present in double dose produces another effect and the chicks become non-viable and die. When present in single dose the unchanged gene, in some way, prevents this happening. The homozygous dominant white can be bred but not reared.

Another characteristic that appears in the dominant white but not in the recessive white, is the deposition of isolated areas of lipochrome colouring. These normally appear in the flights, wing butts and neck areas, their intensity and distribution differing from bird to bird, and becoming more apparent on non-frosted specimens. This would suggest that although the mutation is effective in preventing the unchanged gene from expressing colour, it is not fully dominant. Reports periodically appear of dominant white birds turning pink having been fed carotene. What has happened is not that the white feather has coloured, this we know not to be possible. As stated, the mutated gene is partially dominant to its unchanged lipochrome-producing partner and the degree of dominance varies from bird to bird. Being normally yellow, if the lipochrome colouring is weak, when mingled with white it will not be obvious unless the bird is examined very closely. When carotene is fed to the bird the pale yellow will change to orange, and the pale orange of a diet-restricted red ground bird will change to red. The overlap of the white then presents the visually pink bird. Two points must be emphasised, however, firstly, that not all dominant white birds will express colour when given carotene in their diet, in fact the greater majority will not. The degree of dominance of the factor will determine the presence or lack of lipochrome colouring, the yellow in the flights being the exception, as this is almost always obvious. Secondly, the change can only happen when the bird moults. Once formed, the feathers will not change colour, no matter what is fed.

The exhibition specimen resembles its recessive counterpart, with points being deducted for increased presence of lipochrome.

It has been a long-held fallacy among breeders of type canaries that a white bird must always be paired with a frosted yellow mate. This is, of course, ridiculous, the white factors being responsible for the restriction of lipochrome colouring in the feather not in its structure. Although more difficult to distinguish, both frosted and non-frosted feathers are quite common and by studying them we can ensure that normal frosted × non-frosted pairings are effected.

Self Canaries

The Green

In all essential characteristics the domesticated canary in its normal self form resembles the Wild Canary (*Serinus canarius*) from which it is derived, except that the latter is of a more grey colour. The term 'green' is in fact a misnomer, in that neither the wild canary nor its domesticated counterpart possesses any green pigment. The green coloration is an optical illusion caused by the superimposition of the brown and black melanins on a yellow ground colour.

It had been thought until recent times that there were only two forms of melanistic pigment. Firstly, the brown edging to the feather, known as 'phaeomelanin brown', and, secondly, the black pigmentation which is located on each side of the centre shaft, which totally covers the whole feather on its lower third and which is called 'eumelanin black'. Recently a hypothesis has been put forward that within this black pigmentation either intermingled, or separately deposited, but masked by the black, exists a third melanistic pigment, coloured brown, and called 'eumelanin brown'. This hypothesis has been given added weight by some of the results of the experiments conducted on the satinette factor, but we must remain partly sceptical until its existence has been proved beyond doubt.

Very few greens are exhibited today in their own right (this does not, of course, apply to the red orange and rose ground equivalents). Breeders prefer to keep these birds for breeding because their stamina and clearly defined variegation patterns make them invaluable as outcrosses for the production of other mutations. If exhibited, however, the bird should be of an even grass-green colour with dark beak, legs and feet, and the striations distinct.

For the breeders of self greens there is only one ideal pairing, viz.:

self green × self green (preferably homozygous).

This will produce all greens, mostly self but some will show either one or two tail feathers of a clear appearance, this we term as 'foul'.

The term 'green' is now used only to describe a yellow ground bird with the black and brown melanins superimposed upon it. Where changes have occurred in the ground colour the terminology changes.

1 Dominant and recessive white ground become silver blue.
2 Red orange ground becomes red orange bronze (originally red orange green).
3 Rose ground becomes rose bronze.

Where the ivory factor is present in the gold and silver series, the word 'ivory' is simply added to the relevant name. Hence a green, where the ground colour is affected by the ivory mutation becomes an ivory green and the white equivalent an ivory silver blue.

The only genetical difference between any of these birds is in their basic ground colour. The genes responsible for the black and brown markings being identical.

Although we have discussed only the green, the same principles apply to all other birds of the same melanistic genetical make-up, but obviously other considerations have to be made to ensure that the correct hue in the ground colour is achieved.

The Brown

The brown (cinnamon) canary appeared as the result of a spontaneous change (mutation) in the gene responsible for the development of eumelanin black. Whether this happened in a domesticated or wild canary is not known, and is of no great importance. It was, we know, being cultivated in France as early as 1709.

The brown is a recessive sex-linked character, therefore the passage of the gene producing brown is determined by the sex chromosome, designated the 'X chromosome'.

The effect of the mutation is to change the plumage coloration, that is black in the normal green, to brown. It also changes the eye colour from black to pink, and the dark legs, feet and beak to a flesh colour. The description 'pink' when describing the eye colour of the brown is somewhat misleading, but it is used universally by canary-breeders. On hatching, the eyes of a brown chick appear a dark red colour, this colour darkens as the bird grows and becomes difficult to distinguish in an adult. The coloration in the plumage is known as 'eumelanin brown', but whether it is of the same composition as the existing brown pigment thought to be present as the third melanin in the normal green canary has yet to be determined.

As the brown gene is situated on the sex chromosome a male bird can be one of three types:

1 A non brown (green).
2 A brown carrier (green with one normal allelomorph and the other mutated).
3 A brown.

With 1 and 2, the phenotype is the same. The third bird has two brown genes, one inherited from the father and the other from the mother.

The female possesses only one sex chromosome, the other (designated 'Y') carries no genes, and the bird must, therefore, be either a brown or a green.

There are five possible matings that can be considered when studying the inheritance of brown, these are as follows:

	Parents		*Young*	
	Father	*Mother*	*Sons*	*Daughters*
Pair 1	non brown (green)	brown	green/ brown	green
Pair 2	brown	non brown (green)	green/ brown	brown
Pair 3	non brown carrying brown (green/ brown)	brown	50% green/ brown 50% brown	50% brown 50% green
Pair 4	non brown carrying brown (green/ brown)	non brown (green)	50% green 50% green/ brown	50% brown 50% green
Pair 5	brown	brown	brown	brown

Note: The percentages shown are theoretical.

Using the letter B to designate brown the chart appears as follows using a genetical inheritance table.

	Parents		*Young*	
	Father	*Mother*	*Sons*	*Daughters*
Pair 1	XX	XB Y	XB X or XXB	XY
Pair 2	XBXB	XY	XBX or XXB	XBY
Pair 3	XBX or XXB	XBY	XBX or XXB or XBXB	XBY XY
Pair 4	XBX or XXB	XY	XBX or XXB or XX	XBY XY
Pair 5	XBXB	XBY	XBXB	XBY

This could be illustrated as follows using the following symbols:

Z = black, z = absence of black (brown), O = oxidation factor.

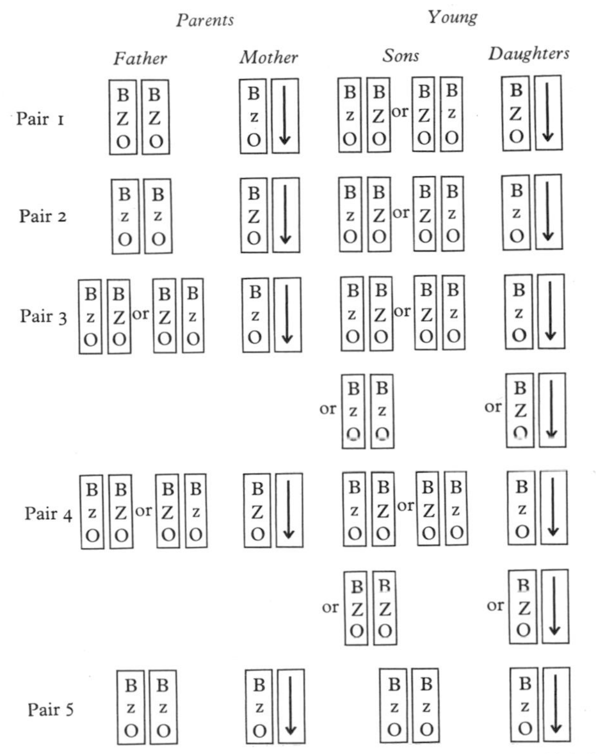

It has long been advocated that the breeder of exhibition browns should always pair self brown to self brown. But if we study any of the recognisable mutations now known and bred, we observe that after two or three years of pairing a mutated specimen to its like, both size, colour, and also in the self series, deposition of melanin gradually start to diminish. This can easily be rectified by backcrossing to a bird with the phenotype of a green, even though possibly its genotype is different from the normal green. We must review the advice of our predecessors, and now say that, ideally, self brown should be paired to self brown to produce exhibition birds but every third year, at least, a green bird should be introduced into the breeding stock. The ideal obviously is a green male carrying brown.

Unlike the green canary, the brown keeps its name regardless of its ground colour. Thus, a yellow ground bird with brown melanins is a gold brown, a white ground bird is a silver brown (previously known as a 'fawn'), a red orange ground bird is a red orange brown, and a rose ground bird is a rose brown. The term 'ivory' is added, when applicable, as with the green.

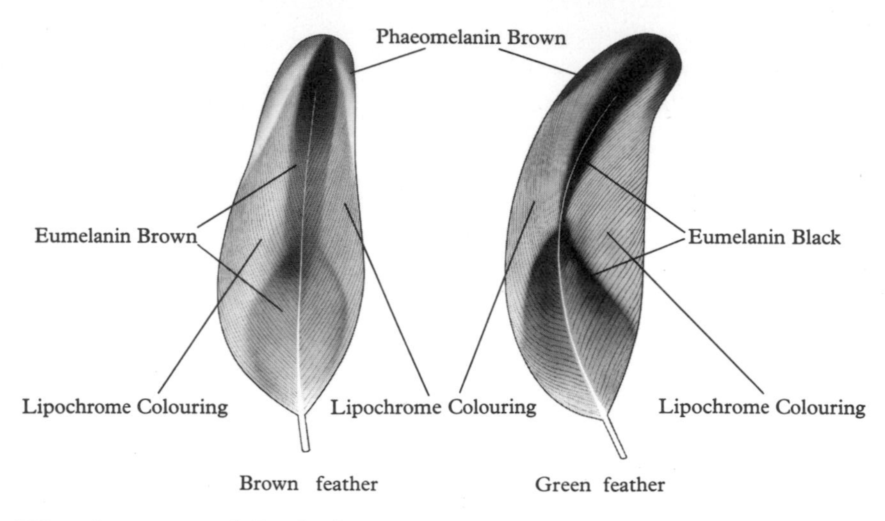

The Agate and Isabel

The common conception of a coloured canary is a red ground bird. If, however, we ignore the brown, the mutation that started general interest in the breeding of canaries for colour rather than for shape or song was the appearance in Holland about 1900 of an ash-grey youngster from a pair of greens. A bird similar in appearance was recorded in France in 1709. This bird was referred to as 'serin agate commun' and in view of this, the name given to the new mutation was the 'agate'.

The factor proved to be a sex-linked recessive that occurred on the gene responsible for the production of melanin. This gene operates quite independently from the genes for the production of lipochrome colour and the mutation has no effect on the ground colour. The agate factor acts as a modifier of melanin causing a bird's dark pigments to become paler, including the colour of the underfeather, which changes to a dark grey colour. The legs, feet and beak also lose most of the black coloration found in the green. Although the lipochrome is not directly affected by the agate mutation, the effect caused by the dilution in density of melanistic pigment allows more ground colour to be exposed, and consequently an agate often appears to be deeper in colour than a normal green. This applies equally to the brown version known as 'isabel'. This is the only mutation where we differentiate between the green and brown versions by giving them different names. The factor used to be known as the 'dilute' and this term was prefixed to green and brown to describe their mutated versions.

With all sex-linked recessive mutations, a female having a sole sex chromosome can appear in two forms only, i.e. with the normal allelomorph or with the mutated gene. In the former she will appear normal, while in the latter she will show the full effect of the mutation. A male on the other hand has two sex chromosomes and can either possess a pair of normal genes, or one normal allelomorph plus one mutated gene, or alternatively a pair of mutated genes. In the first two instances, although having different genotypes, the phenotypes will be identical. In the third instance the phenotype will change to show the full effect of the mutation. The first bird is referred to as a 'normal', the second as a 'normal carrying', or 'split' for the mutation (in this instance agate) and the third will be an agate.

There are five possible pairings to produce agates:

1. Agate male × normal female which produces normal males carrying agate and agate females.
2. Normal male × agate female which produces all normal young, the males carrying agate.
3. Normal male carrying agate × agate female which produces normal carrying agate and agate males and normal and agate females.
4. Normal male carrying agate × normal female which produces normal males and normal males carrying agate and normal and agate females.
5. Agate male × agate female which produces all agate young.

To transfer the agate gene to the brown, to produce isabels, requires a crossover of the relative gene during separation of the chromosomes on the formation of the germ cells. This is covered fully in the section on inheritance. By following the normally expected inheritance patterns we can see that the production of an isabel from an agate male × brown female pairing will not be achieved. It also follows that a green male carrying agate and brown paired with either a brown or an agate female, will not produce the desired result. This may be illustrated as follows.

Fig. 1

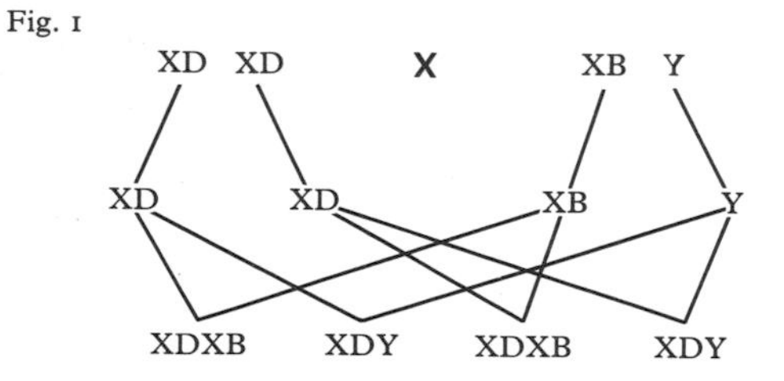

Note: The letter D is used to denote the mutated gene with B as brown.

From this we can see that in Fig. 1 an agate male when paired to a brown female may donate one of his X chromosomes (on which the agate factor is present) to the X chromosome of the female to produce a young male with one mutated gene on the X chromosome and with one brown-producing gene on the other. When joined with the Y chromosome of the female, however, an agate female must be produced.

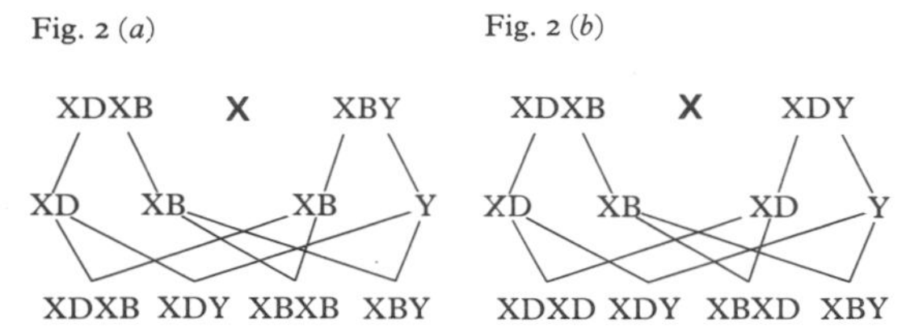

Figure 2 (*a*). The X chromosome of the male carrying the agate gene, when joined with either the X or the Y chromosome of the female will give identical results to the original pairing. The X chromosome of the male carrying the gene for brown, if joined to the X chromosome of the female, will produce a homozygous brown male, or alternatively will produce a brown female if joined to the Y chromosome.

Figure 2 (*b*). In this example the male's X chromosome carrying the agate factor can either produce an agate male or an agate female, depending upon whether it joins with the X or the Y chromosome of the female. The X chromosome of the male carrying the brown factor when joined with the female's X chromosome will produce a green male carrying both brown and agate, or will produce a brown female if it joins with the Y chromosome.

At no time can an isabel be produced unless the chromosomes of the male exchange genes on separation so that one of the X chromosomes carries both mutated genes.

Having achieved this crossover the mutation appeared in two forms, both of which were yellow ground birds. The agate appeared as described earlier, as a yellow bird on which the normal black and brown markings had been diluted giving an ash-grey colour. The gold isabel differs from the brown version in that all depositions of melanin are diluted in intensity giving a much more delicate effect. The colour of the underfeather is also diluted to a dark beige colour. Following the appearance of the two white mutations and then the red and its mutated version the rose, the agates and isabels were introduced into the various self versions to produce a large variety of different colours.

Today the agate and isabel with their normal counterparts, the green and brown, form a quartet known as the 'classic self varieties', and it is to these varieties that all succeeding mutations are normally added.

The ideal agate or isabel is a paler version of the equivalent normal self and like every other canary, whether a mutated version or not, some specimens are superior to others. Although the factor for the production of melanin has mutated, the other factors determining intensity and distribution of melanins may not have done. So, in some instances, it is difficult to see any real difference, except by a very close examination, between a normal and a mutated specimen. In other instances the bird will appear almost devoid of pigment. This is the ideal providing that all striations are distinct, particularly in the flanks, where (especially in the isabel series) a complete disappearance of these markings is not uncommon. To ensure that this does not happen a crossback to a normal is advised every third year at least. In this instance the word 'normal' can refer to either a brown or an agate in the case of the isabel. The agate version does not have the same tendency to lose size or striations as the isabel and backcrosses are needed less regularly. This principle applies to all ground colours.

With the introduction of the opal factor in the 1950s quickly followed by the pastel, ino and satinette mutations, the breeding of all the classic varieties as a separate and distinct entity declined, the birds being used only as agents for transmitting these later factors. Gradually they are returning to favour particularly in the red and rose series, where the contrast of red, pink, black and brown combinations give particularly beautiful effects. More and more breeders are now attempting to produce once again top-class specimens suitable for the showbench. This must be considered a good omen for the future ensuring that these delightful birds will not follow the fate of some other varieties of canary which have become extinct and perhaps lost to us for ever.

As with most other self birds, the words 'red orange', 'rose', 'gold' or 'silver' precede the words 'agate' or 'isabel' to denote the appropriate ground colour.

The Opal

It is difficult to determine the reason for the fact that it was not until about ten years after its discovery that the opal mutation found deserving popularity. In 1949, a lime green bird with a different striation effect was observed amongst the progeny of a pair of green roller canaries in Germany. The name given to the bird at that time was the 'recessive agate'. On reaching Holland several years later, however, its name was changed to the 'opal'.

The factor proved to be a homozygous recessive but not sex linked, which could account for some of the delay in its general acceptance. This was the first mutation having this inheritance pattern known to emerge for thirty years, the previous example being the recessive white. To breed any mutation that is recessive and not situated on the sex chromosome it is necessary for both parents to have at least one of the genes of the appropriate pair in a mutated state. Invariably the phenotype of both carrier and non-carrier of the mutation is identical, and a planned series of test matings is, therefore, required both to determine the genotype and also to establish the mutation. This process requires much patience and dedication on the part of the breeder, with experimental pairings being necessary to cover every eventuality and to establish conclusively the genotype of every bird used. Sometimes it is posssible that the mutation will establish itself very quickly, but more often than not many years of work are necessary before free breeding strains are established.

With this in mind one can imagine that many such mutations have occurred in the past and also exist today, but have never become visual. If this is the case it is quite feasible that some new forms will make their appearance in the years ahead.

Initially, when it appeared, it was necessary for the opal to be paired back to a normal, unless of course the parents survived, but this fact is not recorded. If the latter pairing was possible, the young produced would all have been visual normals but would all have carried the factor for producing opal. These birds could then have been inter-mated or, alternatively, paired back again to the opal to produce pure visual opal birds. Now that the species is well established the following pairings can be used:

1. Opal male × opal female will produce all opal young.
2. Opal male × normal female or normal male × opal female, will produce all normal young carrying the opal factor.
3. Normal male carrying opal × opal female, or opal male × normal female carrying opal, will produce opal and normal carrying opal young.
4. Normal male carrying opal × normal female, or normal male × normal female carrying opal, will produce all normal young, 50 per cent of which will carry the opal factor.

The above percentages are theoretical expectations.

The effects of the mutation are numerous. Firstly, it obliterates almost all trace of the phaeomelanin brown located at the edge of the feather of both brown and green birds; secondly, it obliterates completely the eumelanin brown in the striations of the brown bird, although only reducing the brown of the underfeather; thirdly, it alters the colour of the eumelanin black striations of the green bird to a dark silvery grey. But again it does not have so powerful a reducing effect on the colour of the underfeather. The fourth effect is most unwelcome because in the green variety, changes take place in the composition of the feather causing particularly the secondary flight feathers to often lie incorrectly. Also the body feathers may curl in a manner not unlike those of frilled canaries. When the opal factor is introduced to any of the other three classic canaries, however, the fault is almost always corrected. The amount of poor feathering in the green opal can be partially controlled by using a normal bird carrying the factor as one of a pair, particularly when the bird is either a brown or an isabel. When a green opal is paired with a green opal the feather fault is accentuated, and in consequence this pairing cannot be recommended. Except for the general rule advising backcrossing to a normal at least in every third year as in other mutations, any of the classic varieties in opal form other than the green can be safely paired together. Because of the known feather faults in opal birds, however, the general rule of pairing frosted bird to non-frosted bird should be followed whenever possible. Under no circumstances should a non-frosted × non-frosted mating be attempted.

If the known effects of the opal factor are applied to the classic varieties the results are as follows:

Green opal – all traces of brown are removed leaving only silvery grey striations of the same density and distribution as the classic green. The opal's grey visual effect varies according to the bird's ground colour, whether it be yellow, white or red, or their mutated versions. As with all varieties the exhibition specimen needs to have a bright, deep ground colour, with the striations distinct and clear. The feather quality should also be perfect and the beak, legs and feet very black.

Agate opal – all traces of brown are again removed, leaving a diluted version of the green opal. As these birds do not suffer from the curled feather fault they are kept by more breeders and are, therefore, more in evidence on the showbench, particularly in the mutated red orange (rose) variety. The effect of the neat silvery grey striations on a pick background gives a most pleasing delicate appearance as it also does on the gold and silver agate opals. The depth of ground colour in the red orange version, as with all birds displaying delicately coloured melanins, tends to mask the silvery striations and much of their delicacy is lost.

Brown opal – in most examples, unless one studies the bird very closely it can easily be mistaken for a clear. Occasionally deposits of brown are left in the flight feathers of the wing which gives an indication that the bird is a mutated self variety rather than being a non-pigmented bird. More often than not, however, it is necessary to study the underfeather to be absolutely certain.

Isabel opal – as with the brown the visual appearance is of a clear bird. In poor light, so pale is the pigmentation even in the underfeather, that one could easily form the opinion that the bird being examined was non-pigmented.

By virtue of the relatively uninteresting appearance of the brown and isabel opal to the breeder of self species, and because of the problem of curled feathers in the green opal it follows that the agate opal is the bird that is most often exhibited. This apart, however, the other three varieties should certainly not be discarded by the breeder of opal canaries. The brown and isabel are invaluable as correctors of the curled feather fault in the green opal. Also, because of the difference in coloration of the eumelanin black pigment in the green opal, that variety is well worth persevering with in spite of the problems which might arise.

The Pastel

Approximately ten years elapsed between the discovery of the opal mutation and the appearance of the next one that affected the melanins. In or about 1960 a Dutch breeder named Kollen produced from a pair of red orange isabels a female which appeared to be totally devoid of melanins. Although some experimental breeders had been attempting to do just this with classic isabels, so unexpected was the appearance of this bird that it was apparent that a spontaneous change had occurred on the melanin-producing gene. When moulted, the bird, which was frosted, showed a suffusion of brown in place of the striations which would normally be expected. From that day to this, the factor has never been fully exploited nor totally defined. We do have some facts which can be listed but many questions still need to be answered.

The factor is a sex-linked recessive and, therefore, needs to be present on both genes of the X chromosome of the male to be visual. A female having only one X chromosome is either a normal or a mutant.

Few problems exist in the breeding of isabel or brown pastels, and the identity of the bird is usually obvious from examination. It is different, however, in the agate and green series, but before following this controversial trail let us return to the isabel and brown.

The isabel pastel is visually almost a clear bird, i.e. one in which both suppression of melanin factors are present, thus preventing deposition of melanin. The underfeather colour is pale beige. The eumelanin brown located at the centre of the feather and already diluted by the isabel factor now disappears, leaving only the phaeomelanin brown, and that also is further reduced. This gives an over-all brown suffusion across the bird's back and is more noticeable in the frosted version (there being a greater width of feather on which phaeomelanin brown is deposited) than in the non-frosted version. It is often said that the pastel factor also adversely affects the lipochrome colour, but when this is apparent it may well be that the bird would have been inferior in colour anyway. When the pastel factor is combined with the dimorphic isabel there is a most striking effect. The bird to a great extent resembles a clear dimorphic but with beige replacing the normal white colouring.

In the brown series faint striations of eumelanin brown can usually be distinguished in addition to the brown suffusion. The qualification 'usually' is used because, as with all self varieties, other factors are present which affect to some extent the determination of density and distribution of melanin. In some instances the eumelanin brown is almost totally absent, while in others it is very apparent.

Show specimens of these birds should always be brightly coloured regardless of ground colour, with the isabel pastel showing a minimum of brown suffusion. In other words it should appear as a first-class frosted clear with brown replacing the white frosting effect found in clear birds. The brown version should show clear and distinct striations with maximum dilution of density coupled with the reduced suffusion as described in the isabel pastel.

As with all sex-linked mutations, the inheritance table is quite straightforward, and the brown and isabel specimens can be intermated to give the expected offspring, without the one adversely affecting the other. The five pairings to produce the pastel are, therefore, as follows. In each instance, isabel birds are used to simplify the situation:

1 Isabel pastel male × isabel pastel female will produce all isabel pastel young;
2 Isabel pastel male × isabel female will produce isabel males carrying pastel and isabel pastel females;
3 Isabel male × isabel pastel female will produce isabel males carrying pastel and isabel females;
4 Isabel male carrying pastel × isabel pastel female can produce isabel males carrying pastel or isabel pastel males and isabel pastel or isabel females;
5 Isabel male carrying pastel × isabel female can produce isabel males or isabel males carrying pastel, and isabel or isabel pastel females.

To introduce brown is equally straightforward, but we must remember that two factors, both sex linked, are now involved. To give an example of this we will use an isabel pastel male with a brown female, and following on from this, one of the brown males carrying isabel and pastel with an isabel pastel female.

1 Isabel pastel male × brown female will produce brown males carrying isabel and pastel, and isabel pastel females.
2 Brown male carrying isabel and pastel × isabel pastel female can produce:
   (*a*) Males
      Brown carrying isabel and pastel
      or brown pastel carrying isabel
      or isabel carrying pastel
      or isabel pastel, and
   (*b*) Females
      Brown
      or isabel
      or brown pastel
      or isabel pastel.

To transfer the pastel factor from the isabel and brown series to the agate and green series requires a crossover of genes, and although this proved difficult it was eventually achieved, thus making the mutation available in two further forms.

Fortunately, or unfortunately (depending upon one's approach to the subject) the identification patterns of both the agate and the green pastels did not conform to the recognisable patterns of the isabel and the brown. It is, therefore, necessary to study each of them quite independently and then to try to describe both the ideal bird and the one normally found.

The agate pastel was, at its inception, confused with the classic isabel, and in many cases this confusion still exists. The phaeomelanin brown, as with the isabel pastel, remains virtually intact, but the eumelanin black takes a variety of forms. In most instances the width of melanin deposit in the feather is slightly reduced resulting in the visual colour changing from black to charcoal grey. Whether this is due to the factor reducing the eumelanin brown that is thought to be intermingled with the eumelanin black, or whether it is due to a genuine change in composition is still being debated. In some instances the striations appear broken, in others they appear continuous as in a classic agate. Rarely do they appear distinct, in fact it would seem that the only time they do so is when the optical blue (reduction of brown) factor is also present. In this instance the suffused brown is greatly reduced or may even be obliterated completely leaving very clear and distinct grey striations. This would seem to indicate that the smudged effect

Frosted Green Opal

Frosted Gold Agate Opal

Recessive White Blue Opal

Recessive White Agate Opal

Silver Isabel Opal

Frosted Ivory Gold Agate Opal

Frosted Gold Brown Pastel

Non-Frosted Gold Isabel Pastel

Recessive White Brown Pastel

Recessive White Isabel Pastel

Frosted Gold Agate Pastel

Non-Frosted Rose Agate Pastel

Rose Greywing

Rose Greywing (wing and tail features)

Frosted Red Orange Brown Ino

Dimorphic Red Orange Brown Ino

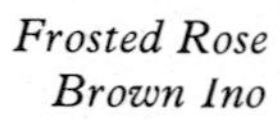

Frosted Rose Brown Ino

Frosted Gold Brown Ino

Frosted Ivory Gold Green Ino

Frosted Ivory Gold Isabel Ino

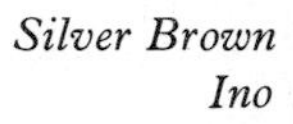
Silver Brown Ino

Silver Isabel Ino

Dimorphic Red Orange Isabel Satinette

Frosted Rose Brown Satinette

Frosted Gold Green Satinette

Frosted Gold Brown Satinette

Frosted Ivory Gold Isabel Satinette

Frosted Ivory Gold Agate Satinette

Ivory Silver Isabel Satinette

Recessive White Isabel Satinette

Silver Brown Satinette

Recessive White Agate Satinette

is caused not by any alteration to the eumelanin black but by the masking effect of the phaeomelanin brown. The colour of the underfeather also varies greatly, ranging from the almost black colour of the classic agate to a dark silvery grey. This underflue variation does not always coincide with a corresponding depth in the colour or extent of the striations, in fact the opposite is more often the case. The ground colour also seems to be affected by this factor in many ways. When the red version is showing a lot of brown suffusion the ground colour seems to be much deeper and brighter. With the paler rose, yellow and white equivalents, when there is a high proportion of brown suffusion the ground colour invariably appears duller. In many species of bird where sexual dimorphism occurs, the female often possesses more brown within its composition – the difference between the male and female Blackbird (*Turdus merula*) is a good example of this – and it is so with the hen canary. In consequence it is almost unknown to see a first-class specimen of an agate pastel female – if we are to compare females with males. Perhaps as our knowledge improves we will become more sophisticated and will make separate standards of excellence for males and females.

The agate pastel makes a complicated study but the green version should certainly satisfy the most enquiring mind as it is even more complex. The phenotype can vary from a bird almost resembling the classic green to one exhibiting the full 'greywing' effect. The latter is the accepted ideal bird and, needless to say, is the most difficult to achieve. Notwithstanding the density and distribution of melanins, to show the greywing characteristic the bird must be completely homozygous for this factor alone. A green pastel carrying other factors will not give the desired effect to the same extent.

A heterozygous green pastel can appear in various forms from almost the classic green to a bird sometimes described as a 'lacewing', but can never appear as a greywing. Unfortunately, it is just not possible to say, for example, that a green pastel carrying the brown factor will give a certain effect while a green pastel carrying the agate factor will give another effect. One of the strange characteristics of the mutation, however, is that whatever its genotype the green pastel always has a silvery grey underfeather that is paler than any of the agate versions.

The striations on the back and flanks of the green pastel seem to be less affected by this factor than those on the flights and tail where extreme variations exist. On the back, even in birds that appear almost as classic greens, the eumelanin black changes colour to a certain degree resulting in a charcoal grey effect, as in the agate pastel. In the case of the classic green, the wing and tail feathers appear black, but in the green pastel they fluctuate from a similar shade to the full greywing effect in which the black colour is 'pushed' into a 3 millimetre bar on the tip of the feather, the remaining part of the feather changing to a silvery grey colour. In between these extremes there appears a multitude of variations including the lacewing effect in which a series of grey bars extend across the wings and tail. This is probably the most common male version of the factor. Whenever a partial or full greywing effect is present it is invariably most accentuated before the bird's first full moult, i.e. in the second year, after which the tendency is for the pattern to revert to some extent to the normal.

It has already been stated that in order to produce the full greywing effect it is necessary for the bird to be truly homozygous for the pastel factor. A female never expresses the factor in the same way as the male because she can only receive a single dose having only one X chromosome. Although the inheritance pattern is identical to the isabel and brown series, bearing in mind that the original pastel was a doubly mutated version with the crossover occurring to produce the green and agate versions, it is still very much a matter of luck to produce a homozygous green pastel. We now know that this is the only form in which the greywing will appear and as this is a comparatively new discovery the majority of our breeding stock is still heterozygous, and only by careful selection will a free breeding strain of greywings be eventually obtained.

To be a show specimen the agate pastel must show distinct even striations with the brown reduced to a minimum. The full effect of the greywing should be evident in the green pastel. As the latter phenomenon can only be fully appreciated visually by spreading out the feathers in the wings and tail it is more a breeder's bird than an exhibitor's. To be most successful on the showbench, therefore, a breeder should perhaps concentrate on the isabel, brown, and agate versions, leaving the green pastel to the experimenter until more fully developed and understood.

Each of the mutations with which we are concerned produces a different effect either on the ground colour or on the coloration or distribution of the melanins. As all melanistic mutations produce a different effect it is, therefore, not normal when attempting to breed exhibition specimens that one or more of these mutations be joined with another. Classes are usually scheduled at exhibitions so that a mutation expressed in any of the four classic varieties can be entered. It is unusual for classes to be provided for birds that have a double mutated phenotype.

The pastel, when linked with any of the other mutations in the brown or isabel series, produces a visually non-pigmented bird. This is normally contrary to that for which we are searching, and in consequence it cannot be recommended where a non-pastel phenotype is required.

Experimental breeders may wish to explore the full possibilities of introducing the pastel factor to other mutations in the green and agate series. Little is recorded of the results of such experiments and breeders are advised to proceed carefully and to keep full details of all colour combinations produced and expected.

The Ino

To attempt to accurately define the effect that the ino factor has on melanistic pigment is difficult, the reason being that the factor appears to manifest itself in at least three forms. When the mutation originally occurred it was in the progeny of a pair of classic red orange isabels. The phenotype of this version could easily be confused with that of a clear red, or a red orange isabel pastel with very little suffusion of brown. It possessed, however, bright red eyes that did not alter in colour as the bird grew older. Also, inspection of the underfeather showed it to be a pale beige, suggesting the existence of another mutation. At first it was thought that the new factor, in addition to giving the bright red eye, prevented both eumelanin and phaeomelanin brown from appearing, when introduced to classic brown stock. However, this theory was proved incorrect. The first type of brown inos show white coloration where the eumelanin brown normally exists, i.e. down the centre of the feather. Sometimes this white area extends to the tip of the feather. In other cases it does not, the edges of the feather being brown, although the area of distribution and density, as might be expected, differed from bird to bird. In some instances the brown pigmentation covers most of the feather leaving only a narrow strip of white down the centre, whereas in others the brown is restricted to the edge. Lipochrome colouring is situated between the deposit of the brown and the white. The brown pigmentation in the second type of inos follows similar patterns except that the lipochrome extends between the brown on either edge giving a solid colour throughout the central area of the feather, the white disappearing completely.

The third version gives a total reversal of these patterns, with the brown pigment being situated down the centre and on the tip of the feather and the lipochrome colouring on the remaining parts.

In each of these three types the underfeather is identical in colour, being a pale brown.

To try to explain the reason for the existence of these different versions without the support of intensive experimentation would not be sensible. Obviously what is needed are group experiments to try to find out whether the different patterns breed true to type, or whether there are variations in the phenotype of the progeny from each of the three. What also needs to be fully researched is the identification of the composition of the brown coloration when it appears on the edges of the feather. Certainly it would seem to be too intense and widespread to be positively identified as phaeomelanin brown. The most likely explanation is that it is eumelanin brown that has been displaced from the central area of the feather. It is possible, although improbable, that the third version is the result of another mutation that, because of its eye colour, was assumed to be an ino when it first appeared.

The green series inos also follow these three patterns, but in the first and second versions rarely is the brown pigmentation so well defined or distributed. A secondary consequence resulting from experiments to determine the composition of the brown edging is that if it is proved to be displaced eumelanin brown, then this could possibly confirm the presence of a third melanistic pigment in green series birds. A question that must follow from these experiments with green inos is this – does the mutated gene cause the eumelanin black to appear as brown when the bird's phenotype follows the third pattern, or does it prevent it appearing, thus leaving only deposits of eumelanin brown?

In the green ino the colour of the underfeather changes from the normal black to a dark grey and this is found in all three versions.

As with the isabel ino, the agate usually appears to be non-pigmented but carries a pale grey underfeather. In some instances, however, both isabel and agate inos show very faint traces of melanistic deposition. This indicates that even when the mutated diluting gene of the agate and isabel is combined with the mutated ino gene, if a bird is high in density of brown, be it phaeomelanin brown or eumelanin brown, they are not totally dominant over the unmutated genes.

The eye coloration, while usually very bright red, does tend to vary from bird to bird with no definite pattern emerging at present. It has been suggested that it is usual for the isabel, particularly, and also for the brown versions to have a brighter eye than the agate or green birds, but insufficient records exist for this to be proved. When first hatched an ino chick is easily identified because of the eye colour, the whole area round the eye appearing almost transparent. Very rarely can it be confused with the plum red colour found in newly hatched brown birds.

Whatever version prevails all are homozygous recessives and, therefore, follow the normal inheritance table one expects. This is as follows:

1 Ino male × ino female will produce all ino young.
2 Ino male × normal female, or normal male × ino female will produce all normal young carrying the ino factor
3 Normal male carrying ino × ino female, or ino male × normal female carrying ino will produce 50 per cent ino young and 50 per cent normals carrying ino.
4 Normal male carrying ino × normal female carrying ino will produce 25 per cent normal, 50 per cent normal carrying the ino factor, and 25 per cent ino young.
5 Normal male carrying ino × normal female, or normal male × normal female carrying ino will produce all normal young 50 per cent of which will carry the ino factor.

All inheritance tables are based on theoretical possibilities. This table is no different from any of the others, and, with the exception of the first two pairings where the genotype of the progeny is guaranteed, all chicks produced may carry the genotype of any one of the possibilities, or they may follow any combination of the possibilities.

Where the word 'normal' is used it denotes a bird with a non-ino genotype. Although the birds generally used are the classic self varieties, breeders in Europe have transferred the ino factor to their lipochrome varieties. The full beauty of the mutation is lost by doing this, but clear birds with bright red eyes are bred. In this instance a bird described as 'normal' would be a clear lipochrome variety with dark eyes.

For either a brown or a green ino to be considered as an exhibition specimen, the brown pigmentation must be as dense as possible. This gives an over-all hammered copper effect in some cases which on a top-quality specimen is not only most pleasing, but is also totally different from any other effect that we have yet created. With the ino factor, the rich ground colour of the red orange versions tends to mask the full effect of the deposition of melanistic pigment, except when the dimorphic pattern is also incorporated. Thus it is normally in the rose, gold and silver versions that the full beauty can be appreciated. One must stress again, however, that with the rose version, loss of size and colour will usually result if a red orange specimen is not introduced as one of a pair in at least every third year.

It is within specimens of the ino mutation, both brown and green versions, that the difference in distribution of the melanins between male and female is most noticeable. Almost never does the male carry the same amount of melanin pigment round the head, chest and flanks, as the female. This, of course, puts them at a disadvantage when competing against females at exhibitions.

Most countries provide only one class for the ino mutation, but in order not to exclude the particular beauty of the male inos from being seen on the showbench some of the larger shows in Britain have classes to accommodate them.

Inos are frequently not so hardy as the other mutations and because of this it is recommended that, certainly in alternate years, one of a pair should be a normal carrying the factor. The greatest problem with rearing inos seems to be that often the youngsters will not raise their heads upright when a parent is trying to feed them. Instead they hold them horizontally, this being particularly true during the first three or four days after hatching. Only the most patient parent will persevere with such chicks, especially if other chicks in the same nest act in the normal manner. A breeder would, therefore, be well advised to check on the young inos' progress and, if necessary, to hand feed for a day or two.

Care should also be taken to ensure that an ino chick is not taken away from a familiar cage or its parents until the fancier is satisfied that the bird is not partially blind. While this fault is less prevalent now, in its early days it was particularly noticeable that many had very weak eyesight. Thus many apparently fully weaned youngsters were transferred to flight cages where they gradually became weaker and eventually died. It appeared on investigation that when the birds were in familiar surroundings, they were initially attracted to the food trays by others in the cage and through instinct descended to this area to feed themselves. When transferred to strange surroundings, however, they were unable to locate the areas where the food-dishes were situated, and consequently starved to death.

Although other mutations create puzzles, the ino, possibly because it has only been readily available for six or seven years, poses more questions than most other varieties, and as such must be a welcome addition to any experimental breeder's stud.

The Satinette

Many homozygous recessive mutations need to be present on both genes of a pair before any visual alteration is observed. The satinette factor is, however, an exception in that in one instance at least, its presence, although in no way fully expressed, can be detected in a normal bird which only carries the factor. This is of major importance to a breeder possessing birds which may possibly carry the factor, i.e. males bred from the mating of a normal bird carrying the factor to a normal female.

It was in the late 1960s that the satinette mutation first occurred, but as with all mutations it took time to establish and to breed sufficient stock to allow birds possessing the factor to be released into general circulation. It was not until 1971 that the first examples reached Britain.

The satinette is a sex-linked recessive, and its effects are threefold. Firstly, like the ino, it carries a brighter red eye than that of the classic brown, which does not darken as the bird grows older. The second effect is that it prevents from appearing the eumelanin black which in the classic green or agate is in the centre of the feather. Thirdly, it prevents the phaeomelanin brown which is located on the tip of the feather of all classic self varieties from expressing itself. Perhaps the most curious phenomenon created by this mutation is that the underfeather of all four of the classic self varieties is identical. This colour is dark beige. One might think that this is of little importance for with many other varieties it is not essential to study the underfeather to determine the nature of the bird under inspection. With the appearance of the satinette mutation, however, the long-lasting and popular belief that the canary carried only two different forms of melanistic pigment was suddenly the subject of much debate.

As the factor prevents the display of both eumelanin black and phaeomelanin brown, it is not unreasonable to assume that the satinette versions of the green and agate would appear as non-pigmented birds with a red eye. In many instances this is the case, but there are exceptions. A bird which by virtue of its pedigree must be a green or agate sometimes appears with faint brown striations.

Much thought was given to the possibility that the mutation caused the eumelanin black to appear in a different-coloured form resulting in a reduction in density. These theories, however, do not really stand up to investigation because it seems that there are just as many such birds that are totally non-pigmented. As an answer to this phenomenon it is now being suggested that within the eumelanin black, either intermingled or separately deposited but masked by the black, exists a third melanistic pigment that is being referred to as 'eumelanin brown'. Within the brown series it is noted that the satinette factor has no effect on the density or distribution of the eumelanin brown and it is, therefore, being conjectured that this third pigment is of a similar or, maybe, identical composition.

Let us then summarise the effects of this mutation as related to the four classic varieties. All versions of the satinette have red eyes.

The brown satinette is a bird that carries the distinct brown striations of similar width and density to the classic brown.

The isabel satinette is not dissimilar to the brown satinette, but the striations show the diluting effect of the isabel mutation and their width is reduced. In both the brown and the isabel satinette the only pigmentation remaining is the eumelanin brown, which appears very distinctly because the phaeomelanin brown being absent is unable to mask its effect.

The green satinette can appear as a non-pigmented bird, or alternatively, can show very faint brown striations. It has been decided by most societies that the latter should be sought as an exhibition specimen. This is based on the grounds that a self canary ideally should show some form of pigmentation, even if in a greatly modified form.

The agate satinette can also appear either as a clear, non-pigmented bird, or as one with very faint brown striations, the latter again being preferred.

Normally one can assume that depth of ground colour, type of feather, etc. will affect the visual result of a mutation of the genes controlling melanistic pigment, and this assumption usually proves correct in the case of the satinette. There are instances, however, where inexplicable variations occur. Very few non-frosted satinettes have been bred at the time of writing and it is, therefore, difficult to pass on any confirmed opinion about these birds. The only one on which comment can be made is the rose form, where it is known that the brown satinette is a bird of particular beauty. Little, however, is known of the effects of the mutation on non-frosted birds with other ground colours.

Whatever ground colour or feather type is used, the agate and green varieties stimulate little interest as exhibition specimens, but they are valuable as one of a pair to help retain the striations, particularly in the flanks of the brown and isabel versions.

Having discounted the green and agate satinette, all the following comments relate to the brown and isabel versions only.

In the red orange ground satinette, the frosted bird carries a white colouring in place of the normal phaeomelanin brown frosting pattern. This helps to highlight the brown striations, but without a doubt, to really exhibit maximum beauty the red orange brown or isabel satinette must also be dimorphic. In a good example, the dimorphic brown or isabel satinette is not dissimilar to a non-pigmented red orange dimorphic, but with neat distinct brown striations superimposed on its back and flanks. The brown satinette in this form seems to show less distinct striations than the isabel satinette, the over-all pattern seeming to be smudged. For that reason the isabel satinette is proving to be the most sought after.

In the mutated red orange (rose) as we have already mentioned, the non-frosted version seems to give the most spectacular effect, and here the width of pigment of the brown seems much more distinct than in the red orange version and gives a more pleasing effect than in the isabel. In the frosted version the paleness of lipochrome colouring is disadvantageous as there is not so much contrast with the melanin. In both the frosted and dimorphic versions, although the striations express themselves fully and the birds are extremely beautiful, the effect is somehow less striking.

As the yellow (gold) and its mutated version, the ivory gold, are in some ways equatable to pink rather than to red, because of the lower density of colour, it is reasonable to assume that the same observations will apply. Certainly in the frosted and dimorphic versions the gold brown satinette is more striking that the gold isabel satinette, but as yet few non-frosted specimens have been produced, and it is, therefore, impossible to comment with any authority on their normally expected appearance.

The dominant white (silver) isabel satinettes, as in the case of the red orange versions, seem to have more distinct striations than their brown equivalents. In the recessive white version this is particularly evident. Unexpectedly, when the ivory factor is added to the normal silver isabel satinette, one is left with a bird that can easily be confused with a well-marked agate satinette. Why this should be is not known, but perhaps in time the answer will be revealed.

With all sex-linked mutations there are five possible pairings, and these are listed below. Where percentages are referred to it must be remembered that these are average expectations.

1 Satinette male × satinette female will produce all satinette young.
2 Satinette male × normal female will produce normal males carrying satinette and satinette females.
3 Normal male × satinette female will produce normal males carrying satinette and normal females.
4 Normal male carrying satinette × satinette female can produce normal males carrying satinette, normal females, satinette males and satinette females.
5 Normal male carrying satinette × normal female can produce normal males, normal males carrying satinette, normal females and satinette females.

On average from the last pairing, one could expect 25 per cent of each type.

Mention was made earlier to the fact that in certain instances the presence of the satinette factor could be identified in normal male birds carrying the factor. Unfortunately, this only occurs in the agate series, but at least this is an improvement on all earlier mutations. The black colour normally present in the primary feathers of the wing of a classic agate male is subtly altered to a dark grey colour when the bird carries the satinette factor. Although not immediately obvious to a person viewing the phenomenon for the first time, on comparison with a normal agate the difference becomes obvious. When a mutation is first made available to all breeders the scarcity of breeding stock means that it is only possible for a breeder to purchase a male carrying the factor as his initial investment. In the case of sex-linked mutations, however, the only females available for breeding will be normals. This means that only possible carrier males will be produced. Normally all of these will need to be test mated to determine whether they carry the factor or not. If the initial bird obtained, carrying the satinette factor is either an isabel or an agate, by mating it with an agate female we know that all agate males must be produced, and because the males carrying the factor have a different phenotype to the males that do not carry the factor, this test-mating process becomes unnecessary as the carrier males will be identifiable and one year's work will be saved.

If one inspects the lutino mutations in Greenfinches (*Carduelis chloris*) it will be observed that the effects are very similar to those of the satinette mutation in canaries, i.e. all traces of black pigmentation disappear, the colour of the underfeather is changed to brown and the colour of the eye is changed from black to bright red.

Not many years prior to the announcement of the appearance of the satinette, continental breeders were purchasing lutino greenfinches in large quantities from British breeders, and thought was given to the possibility that the satinette was, in fact, and F2 or F3 lutino greenfinch hybrid, the recessive factor having been successfully transferred. It is not known whether greenfinch hybrids have been proved fertile, but reliable authority suggests that the suppositions are incorrect. It would be very interesting, however, to try to determine whether the two factors occurring in the separate species are in fact one and the same. To do this we need to mate a satinette male with a lutino greenfinch female. If the factors are the same in both birds and are situated on similar genes, then all the offspring should be red-eyed with the males appearing visually non-pigmented.

Continental breeders have succeeded in introducing the satinette factor in addition to the ino factor, into their lipochrome (clear) varieties. The effect, although not startling, does give a different appearance, particularly in the rose, ivory gold and recessive white varieties, where the red eye is most obvious. As the mutation is of particular interest to breeders of self varieties, it is in such birds that we can see and admire its full effect.

As selective breeding and experiments continue it is certain that this variety will become a firm favourite with breeders everywhere, and in a few years' time many of its secrets will be revealed.

The Pearl

The development of this combination of mutations is very much in its infancy and much of the following should be accepted as hypothetical rather than fact, insufficient depth of information being available to form established truth.

The continental breeders have always been far more prepared to experiment with their canaries than the more conservative British. One bird that is bred only in the pure state in Britain has been used for many years elsewhere to produce a variation of the different mutations. This is the lizard canary. To date, the addition of the various ground colours and mutations has not resulted in the production of a sufficiently different appearance for it to have become popular, although there are many interesting examples of blue, red orange, rose brown and isabel birds. The latest combination of factors, however, has at last produced the potential for a really spectacular bird which must surely capture the interest and imagination of even the most pedantic of coloured canary breeders. This is the satinette lizard, which is now being called the 'pearl'. By combining the two factors plus brown, a red-eyed bird is produced that, in the case of a top-class specimen will show brown spots or hyphen marks where the normal eumelanin black striations appear.

It is a characteristic of the lizard canary that the striations, called 'rowings', generally spread farther round the chest than with other varieties, and thus the brown spot phenomenon is emphasised, giving an all-over spotted effect.

In nest feather the bird is believed to appear as a normal satinette, the spotted effect only appearing when the first moult is completed, and (like the pure lizard) lessening at the second moult.

To date the only specimen examined was of poor quality and it would be difficult to differentiate between it and a normal brown satinette. That it was a pearl was without doubt, however, as it exhibited a broken cap, a feature of many normal lizards. Parts of the bird did carry sufficient of the lizard spangling effect to capture the imagination, and it is hoped that we shall be seeing top-class examples in the not-too-distant future.

But as with everything that is really worth while, a tremendous amount of time, skill, dedication and even luck, will be needed to produce the ideal, a task that coloured canary breeders should find an irresistible challenge.

Let us consider some of the possible pairings and problems to be encountered.

The lizard canary is a homozygous recessive, and the mode of inheritance is, therefore, identical to that of the opal, ino and recessive white factors. The possible pairings to produce lizards are as follows:

1 Lizard male × lizard female will produce all lizard young.
2 Lizard male × normal female, or normal male × lizard female will produce all normal young carrying lizard.
3 Normal male carrying lizard × lizard female, or lizard male × normal female carrying lizard, will produce 50 per cent lizard and 50 per cent normal young carrying lizard.
4 Normal male carrying lizard × normal female carrying lizard, will produce 25 per cent lizard, 50 per cent normal carrying lizard, and 25 per cent normal young.
5 Normal male carrying lizard × normal female, or normal male × normal female carrying lizard, will produce all normal young, 25 per cent of which will carry the lizard factor.

Again it must be emphasised that these are average expectations.

This is fairly straightforward, but problems arise when visually normal offspring carrying the lizard factor are used as one of a pair. The factor is a multi-variegated one and the distinctive variegation pattern is not normally obvious in the first outcross. In this case, assuming a self bird has been paired to a lizard, the youngsters more often than not appear as normal selfs. Variegation patterns go awry in the second generation, however, giving a very high percentage of birds with light feathers and foul tails, even when the youngster is paired back to a self bird. This is one of the major problems that has to be overcome in the production of the pearl as a self canary of any variety, to be acceptable as an exhibition specimen, it must have no clear feathers.

As we have seen from our consideration of the satinette factor, all traces of eumelanin black are obliterated, and as the normal lizard shows mainly this pigment a bronze pearl (normal satinette lizard) would appear as a normal green satinette, i.e. virtually clear, and of no real interest. We may find as we progress with this creation, that it is necessary every three or four years to introduce a bronze pearl into our stock to ensure that we lose neither size nor distribution of the melanins, particularly on the chest and flanks. To achieve our original aim it is necessary to introduce the brown factor to our lizard stock. This can be done as a separate exercise, but as we are also looking for the satinette factor possibly the best way of achieving them both at the same time is to mate a brown satinette male with a lizard female. This will produce brown satinette females carrying lizard, and green males carrying brown satinette and lizard. The first two factors are situated on separate genes on the sex chromosome, whereas the lizard factor is on a gene on one of the somatic chromosomes. Ideally, this mating should be carried out with two pairs so that the young males from one pair can be paired with the young females of the second pair. The theoretical expectations of such a pairing, keeping in mind the three separate mutations and the necessity of a crossover of genes is as below.

Green male carrying brown, satinette and lizard × brown satinette female carrying lizard, can produce:

(*a*) Males

1 Green carrying brown and satinette
2 Green carrying brown, satinette and lizard
3 Green satinette carrying brown
4 Green satinette carrying brown and lizard
5 Green lizard carrying brown and satinette
6 Green satinette lizard carrying brown
7 Brown carrying satinette
8 Brown carrying satinette and lizard
9 Brown satinette
10 Brown satinette carrying lizard
11 Brown lizard carrying satinette
12 Brown satinette lizard

(*b*) Females

1 Green
2 Green carrying lizard
3 Green lizard
4 Green satinette
5 Green satinette carrying lizard
6 Green satinette lizard
7 Brown
8 Brown carrying lizard
9 Brown satinette
10 Brown satinette carrying lizard
11 Brown lizard
12 Brown satinette lizard

From this it can be seen that on average one in twelve of the young males (No. 12) and one in twelve of the young females (No. 12) will be the birds for which we are looking. When we also remember that the chances of them being full selfs are remote, the enormity of the problem presents itself.

Assuming that we have been fortunate in our breeding results, the birds we will keep for the third breeding season will be the brown lizard males carrying satinette, the brown pearl males, the brown lizard females, and the brown pearl females.

From any combination of these pairings we can be fairly sure that we will produce pearls in the third season, having regard to the usual warning relating to average expectations:

1 Brown pearl male × brown pearl female will produce all brown pearl young.
2 Brown pearl male × brown lizard female will produce brown lizard carrying satinette males and brown pearl females.
3 Brown lizard carrying satinette male × brown pearl female will produce brown lizard carrying satinette and brown pearl males, and either brown lizard or brown pearl females.
4 Brown lizard carrying satinette male × brown female will produce brown lizard males, half of which will carry satinette and brown lizard or brown pearl females

This then is the procedure to follow in the attempt to breed the pearl using a red orange, gold or silver brown satinette male in the first instance. Although the lizard carries no genes for producing red, it is thought that by the second outcross the red genes will be dominant enough to give real depth of colour.

We have seen some of the uncertainties that we can expect to meet as we attempt to breed an exhibition specimen of this bird. Now consider the additional problems when adding the recessive white or sex-linked ivory factors to the original. It would be superfluous to present the theoretical expectation tables for these pairings now, but the challenge is obviously with us.

This creation is far too new for anyone to speak authoritatively about all the possibilities, but one cannot help but conjecture on the effect that the non-frosted and dimorphic feathers and also the isabel mutation will have on its appearance.

Without any doubt the particular and peculiar problems presented to us by this development will keep serious colour-breeders busy for many years. Let us hope that the challenge and frustrations will not cause too many enthusiasts to fall by the wayside in the quest for what should ultimately prove to be a spectacular addition to the showbench.

Conclusion

Throughout this book we have studied the various creations and mutations that have resulted in giving us such a huge range of variations in colour within the canary. It should be evident that there are innumerable questions to be answered on many separate subjects. Some of these may be answered in the near future, others perhaps never. Experimental breeders, can, therefore, be assured that their time will be profitably employed seeking these answers for many years to come.

In the future where breeding coloured canaries is concerned, anything is possible. It seems improbable that no new spontaneous changes will take place, although whether the effects of any of these will be great enough to create a noticeably different variation of the ground colour or melanistic pattern, must remain hypothetical.

For many years two particular goals have existed in the minds of coloured canary breeders, neither of which have yet reached fruition. These are the production of a black canary and a blue canary.

The blue canary is possibly the less likely of the two to be achieved. Based on current information, the canary carries no genes capable of producing blue, so one must think of hybridisation as being the alternative. Some form of mutation that modifies existing lipochrome-producing genes cannot be entirely discounted however. Records show only limited experimentation in transferring a blue gene to the canary, the bird having been used was the male Indigo Bunting (*Passerina cyanea*). F1 hybrids, coloured green, are recorded from such a pairing, although whether the experiment stopped at this stage is not known. Perhaps one day in the not-too-distant future breeders will again start these experiments.

While breeders may search for a blue lipochrome colour, the black canary, if it is to exist, will ideally carry a black melanistic feather, probably a modification of the normal green. To be able to freely breed canaries with a black lipochrome would be a great achievement, but consider the possibilities if the melanins alter their appearance. What effect would the melanin-modifying mutations known today have on such coloration? What would be the effect when combined with the various ground colours? What could we expect, if for instance, the dimorphic pattern was added? These questions, of course, are impossible to answer, but maybe we are closer to finding out than is generally realised.

During the 1975 breeding season, a youngster produced by a visually normal pair of Gloster canaries was sufficiently different from the expected phenotype to create a great interest among coloured canary breeders. This bird is one that type canary breeders would refer to as a 'three parts dark buff', i.e. a three-quarter variegated yellow ground bird with a normal frosted feather. The areas where no melanistic pigment are distributed are normal in colour as are the wing and tail feathers. A large area on the chest, flanks and part of the back, however, does appear to to be black. On examination of these feathers, the eumelanin black which is normally present in a green feather is intact but, instead of being located merely in the region of the centre of the feather, it extends farther to the edges from about two-thirds up, giving a totally black tip. Lipochrome is situated as normal but obviously restricted in distribution, owing to the black colouring being so widespread, and all traces of phaeomelanin brown have disappeared. With the overlap of feathers, all lipochrome colouring is masked, giving a solid sooty black effect. The bird in question is thought to be a male, but this has yet to be confirmed. If this unusual effect is the result of a mutated gene, and if the bird is a male, then the mutation logically should be a homozygous recessive and both its parents are carrying the factor (apparently the breeder's entire stud is in some way interbred, so this is possible). The other possibility is that it is a dominant heterozygous mutation, but this is more unlikely as only one gene needs to have mutated, and this being the case, the parent that donated the gene should also have the new phenotype. Should the bird be a female, either of the two previous possibilities can still apply, but also the mutation could be sex linked.

The first problem will be to breed completely self birds possessing the factor so that the full effect can be expressed. When this has been achieved, and a full breeding strain has been created, then a start can be made to transfer the mutation to the various ground colours and to add the melanin-modifying factors. The posssibilities are immense, but before we involve ourselves too deeply in conjecture it must be remembered that periodically other black canaries have appeared, but at their second moult their plumage has reverted to the normal green phenotype. The cause of the black appearance in the birds in the first instance has not yet been established, but certainly it could not be the result of a mutated gene. We can only hope that we do not have a reoccurrence of this reversion in this particular instance.

Attempts to breed a black canary by hybridisation have perhaps been more widespread than in the case of the blue canary, but again little is recorded of the progress made. The bird most often used was the male Bolivian Black Siskin (*Chrysomitris atrata*). Hybrids have been bred from this source but as no further information is available it is reasonable to assume that either the experiments were not extended or the hybrid proved to be sterile.

In Britain in 1975, two male hybrids were produced from a male

Black-headed Siskin (*Spinus notatus*) paired with canary females. The male proved to be extremely vigorous, fertilising all eggs from two females, but only the two male hybrids were reared. Although one of the females was a ticked apricot and the other a self bronze, both hybrids are full self specimens and are practically identical. In both instances the black hood, although obvious, is less dense than that of the father, and as with the F1 Black-hooded Red Siskin hybrids the flight feathers of the wing are identical to those of a green canary. Whether these birds are fertile and if so to what degree the black feathering will persist in the F2, are facts that will shortly be known.

In conclusion, the writer can only apologise for not giving all the answers to every question that might be asked. Not only is this beyond his capabilities, but in a section of the fancy where new problems arise yearly he would suggest that no one is in this happy position. By highlighting areas where knowledge is limited it is hoped to motivate experienced breeders into conducting solitary or preferably group experiments in a search for answers. However, if newcomers to the fancy who, on reading this book, gain the information to assist them initially so that they in turn can contribute to future experiments, then the whole exercise will be a success and the author's efforts worth while.

Management and Breeding Techniques

So variable are the methods used by breeders throughout the world to house, keep and breed their canaries that it is almost an impossible task to give a set of rules which newcomers to the fancy should follow. What can be attempted is to point out some of the possible pitfalls, and to suggest methods to prevent their occurrence, give details of methods found generally successful and then leave it to the individual to adopt or adapt such systems to suit himself.

Birdrooms

Before stock selection and purchasing, it is necessary for a prospective breeder to have somewhere to house his acquisitions. Almost every type of enclosure is used to house canaries, including cellars, attics and spare rooms in dwelling-houses. It is perhaps more common, however, to find canaries established in their own building. This can be made from almost any material, but the most popular is of wooden construction.

Two points that need to be taken into consideration prior to the erection of a wooden building (assuming that a floor is to be fitted) are, firstly, that it should be raised above the ground to a suitable height to enable access. This is to detect or prevent the presence of vermin. Secondly, prior to the floor being laid, the underneath section should be coated with a wood preservative.

Although the larger the breeding-room the better, the cost of erection is without doubt a factor that in many instances has to be taken into account but the size is really not necessarily relevant to good breeding results. When deciding on the site for the building, thought should be given to ensuring that all sides are accessible for repairs, etc., and that the site is not particularly exposed to the extremes of weather conditions.

Before deciding on the position of windows and door it is necessary to make a plan of the proposed layout of cages and indoor flights. When first completing a new breeding-room a fancier might consider he has more room than he will ever need but, as interest grows and more birds of different mutations are introduced, space will soon be at a premium. It is far easier to maximise the space available before erecting the building than to make alterations once it is operational. Daylight is obviously necessary to most living creatures, and canaries are no exception, as much space as possible should then be given to windows and/or roof-lights. Opinions differ on the direction in which windows should face, and this obviously varies from place to place depending upon the direction of prevailing winds, etc. Both of the writer's buildings, through necessity, face south-west and he finds no major problems with this position. Once these points have been considered and the building is finally erected, the exterior should be coated with either wood preservative or hard-wearing exterior paint.

The next task is to ensure that the building has a flow of air. This can easily be achieved by cutting out part of the wood panelling and inserting steel ventilation panels. All buildings should be well ventilated, without being draughty, to ensure the birds are comfortable when introduced.

Thoughts should next be given to the installation of any electrical wiring that may be required. Canaries do not in normal circumstances need either artifical light or heat but sometimes such aids can be used to advantage as will be explained later. If this is intended it is better for the installation to have been completed prior to the introduction of stock. There are many appliances specially manufactured for the breeder of birds, including such things as thermostats to regulate heat and devices to both dim the lights and to automatically switch them on and off at predetermined times. Normal light fittings, heaters and electrical points for use with other appliances should also be considered at this time and fanciers are advised that a qualified electrician should be engaged to organise the connection of the mains supply and the internal wiring required. It is recommended that tubular heaters and fluorescent lighting be used, with the natural daylight tubes being particularly effective.

When the electrical installation has been completed the building should be lined so as to give a flat surface where possible. Plywood or hardboard are the materials most often used for this purpose. The use of insulation material is advised by some breeders and condemned by others, but obviously if its use is intended it should be installed prior to the building being lined.

Another job that should be completed prior to the introduction of stock is the fitting of a wire safety door. This can be a wooden frame covered with wire netting fitted on the inside of the door opening in the opposite direction. Alternatively, it can take the form of a separate wire-netting-covered structure covering the door on the outside. The obvious value of such an arrangement is that the outer door can be left open on hot days with no danger of either the birds escaping or predators gaining access.

Cages, Flights and Accessories

Our building is now ready to be fitted out with cages and/or flights and these can also take many forms. Single, double or treble cages can be either manufactured or purchased and then placed on shelves or stacked in tiers. Alternatively, cages can be built into the building. Let us first consider the free-standing cages.

These can be made of wood or metal with the former being more practicable for the average fancier if he intends to manufacture them himself. It is advantageous to have cages of variable size in which case it is recommended that single cages be made to extend as far as possible along the full length of the building, which can be divided into sections by wooden or metal partitions. Many people use double or treble cages, i.e. cages which allow one or two partitions to be inserted thus effectively making the single into two or three separate cages. These partitions are referred to as 'slides'. Whatever cages are used will make little difference to the birds, but as we gain experience it will become obvious that certain combinations make life considerably easier for the breeder. The actual size of the cages is not too important but ideally it is recommended that the minimum size of each unit, i.e. each section of a double-, treble- or multi-caged unit be approximately 380 millimetres length, 300 millimetres height, and 300 millimetres deep.

Punchbar in loose form can be purchased to manufacture cage fronts but specially manufactured cage fronts in a variety of sizes can normally be readily purchased. These need to be surrounded by a wooden frame into which they fit. It is not difficult to arrange the construction of cages so that the whole front section can be easily removed, and this is advised in order to facilitate cleaning, particularly when it is not intended to use a removable tray.

In some places, special glavanised-steel cages are used, the fronts of which incorporate special doors on which baths and nesting-pan sections can be fitted. These cages are the most practicable, but shortage of distribution facilities make them unavailable to the average breeder which is most unfortunate.

When cages are being built into the birdroom similar procedures apply, but it will, of course, not be necessary to fix back panels to the cages. When this system is adopted it is also comparatively easy to extend them round corners so that valuable space is not wasted. This form of caging can also be constructed to give entry to interior flights, which ideally should reach from floor to ceiling to enable the birds to have upward flight movement which is considered to be very important. It should be mentioned at this time that the bottom cage of a section should stand above the ground on a platform about 100 millimetres high. The height of the top cage is not important. Once the cages are constructed they and the interior of the building should be painted with a lead-free paint. White being the brightest colour available helps to keep the building light, but some breeders prefer to use the same colour paint as is used in their showcages, in the belief that this will assist the young birds' show-training when the time for this procedure occurs. Any type of paint can be used, there being advantages with both of the main types. If gloss paint is used, particularly in the cages, it is more easily washed and does not stain. Many fanciers prefer to use water-soluble emulsion paint into which is mixed a solution of an insect-repelling product. With modern-day vinyl emulsion paints, the amount of staining is less prevelant and the speed of application and drying is of immense benefit.

The use of cages is recommended for the breeding of exhibition canaries but either indoor or outdoor flights can be used if required. The main disadvantage in using aviaries is that, if it is intended to keep more than one pair in them, no control can be exercised over which male mates with which female. These enclosures are, however, most useful and beneficial for the housing of stock at other times of the year. They should be constructed with strong, weather-insulated timbers, over which 10 millimetre wire netting is placed. Where there is a possibility of interference from cats or other predators the use of a second layer of wire netting placed on the inside of the timber is recommended.

When the aviary is not constructed to stand on a concrete platform, a trench should be dug round the bottom member, which should be erected on a brick base and the netting extended at least 1 metre under ground. The trench is then refilled. If the flight is not connected to the building and the birds cannot gain access to the room, one end should be enclosed to afford them some protection from adverse weather conditions. The roofing of this enclosure should, of course, be slightly tilted away from the birdroom. Again a safety door should be fitted.

Having constructed or purchased the cages, there are a number of accessories which are needed. All commercially manufactured cage fronts have head-holes through which the birds feed. Over these can be hung seed-hoppers, and drinking or grit-holding receptacles. Most fronts have two sets of holes and it may be convenient to use one for the seed and the second for the grit receptacle. The water is then offered in a fountain-type drinker. There is a slight risk of birds with closed rings getting their legs trapped in the metal attachment used with some types of fountain drinker, and for this reason the use of a hook-on drinker is advocated in place of the grit receptacle, with the grit being offered in a free-standing pot inside the cage. With the newer type of fountain drinker which clips between the bars this problem does not arise. Perches should be wooden and can be either round, oval or square dowelling or natural twigs, the only two stipulations being that they are firmly fixed and are neither too large nor too small for the bird to stand on them comfortably, a diameter of approximately 10 mm being about the right size. Normally two perches need to be fitted per cage unit. These can be reduced to one per unit when the slides are withdrawn.

The floor covering used varies from breeder to breeder, with paper, sand, sawdust and wood shavings being the substances most common. All have advantages and disadvantages and only by a system of trial and error will the breeder arrive at the substance that suits him best. There are breeders who consider cage hygiene to be low in the list of priorities and there are others who insist upon almost sterile conditions. Each of these types will produce 'facts' to substantiate the opinion that their own method is the ideal. It is difficult to set a time-limit on the frequency of cleaning cages, as this is largely dictated by the number of birds housed in each cage. Let common sense prevail!

Stock selection

Now that we have the facilities to house our stock there arises the question of the selection of birds with which to breed. With the vast number of varieties available it is obvious that only in a very large establishment can a pair of each variety be kept. What is recommended, therefore, is that a prospective breeder should visit a number of early cage-bird exhibitions and closely study the various mutations on show. Having decided on the birds which most appeal to him, the breeder should check to see if there are any fanciers exhibiting these varieties who are winning consistently at the shows. This being so, an approach should be made to enquire if the exhibitor has surplus stock to dispose of, or can recommend someone who has. Often, beginners' tastes will change after two or three years, and it is, therefore, recommended that initial purchases should include both lipochrome and self varieties, three or four varieties being the maximum.

Many beginners intent on breeding exhibition specimens tend to invest firstly in red orange and apricot birds in the belief that they are easy to understand and breed. While their progeny will normally possess the expected phenotype, the belief that they are the easiest birds to breed for the showbench should be discounted immediately. It is in these classes that competition is at its fiercest (particularly in Britain) and the odds against a particular bird of this variety being well placed at a show are considerably higher than with many of the self varieties. Whatever his choice, the newcomer should present himself to the prospective vendor as a beginner and he should ask for advice on the selection of his initial stock. Most experienced breeders will be only too happy to give this advice, and rarely will a beginner find himself deceived.

Some birds offered for sale will not resemble very closely those that may be seen winning on the showbench, but it is quite probable that the vendor having kept very concise records has good reason to believe that the stock will produce good show specimens.

With no experience on which to draw, the beginner must assume that the advice being given is sound, until proved otherwise. Five things that a beginner can check before purchasing stock are, firstly, that the bird is alert and lively when in flight or cage. One that sits around is not fully fit, and is not a recommended purchase. Secondly, the feathers should be tightly held to the body and not fluffed out, which would be an indication that the bird is not fully fit. Thirdly, an ideal pair should usually consist of one bird with the frosted and one bird with the non-frosted feather type. There are exceptions to this rule, but if the point is mentioned to the breeder he will explain the reasons for offering stock that may differ from this. Fourthly, while size is of no particular relevance, the general shape of the bird should conform to the accepted standard, a diagram of which appears below.

The Standard Type

Fifthly, feather quality should also appear silky and not in any way coarse.

All these points having been cleared, one final check should then be made on the bird's general health. This is done by holding the bird's head close to one's ear to determine whether there are any sounds of wheezing. Canaries are susceptible to asthmatical conditions which, while not generally fatal, are obviously to be avoided wherever possible.

Some breeders recommend that canaries are kept in true pairs, while others recommend that a male be paired with two or more females. Both systems are widely used and it must, therefore, be assumed that both have their advantages. We will deal with breeding methods later in this section, but this point needs to be kept in mind by the prospective purchaser.

Theories abound on the number of pairs which should be used by the beginner in his first breeding season. The amount of work involved during breeding can vary considerably, but obviously the more pairings, the more work involved. So until a breeder is sure of his methods and has evolved his own systems, it is better that he has a moderately sized stud. On the other hand the beginner is notably enthusiastic and if there is insufficient work in the management of his stud he may become bored and disillusioned, tending to find jobs to do that are best left undone. The birds then become upset and less liable to reproduce and to rear their young.

A happy medium is the best answer. Also it is recommended that the purchases are made some three or four months before breeding commences, to allow the stock time to adapt to both their different environment and to the management techniques used. This applies particularly to the females.

Breeding

For many people the breeding season is the most important part of the annual cycle as it is the period of actual reproduction. For others there is a continuous process with this period being the climax to the year's efforts. It can be a period of constant delight and pleasure, alternatively it can be one long series of disasters. Canaries like human beings, cannot be counted on to perform all tasks successfully. Some make excellent parents, others are very poor and only by gaining experience will a breeder learn to identify the problems quickly enough to be able to rectify them in good time.

If the birds have been purchased three or four months prior to the breeding season they should be placed in the largest cages possible or preferably in indoor or outdoor flights. As with human beings exercise is very important, and the greater the area in which the bird can fly freely the better it is for its health. At this time no form of artifical heat or light is required unless the building is subject to dampness, in which case heaters thermostatically set at 7–10°C are recommended. Canaries will not be adversely affected by cold weather but will in many instances develop sicknesses if their environment is damp or draughty. The use of a heater will help keep dampness at bay.

The diet at this time should be restricted to a basic seed diet, with additional titbits in the form of modest amounts of sweet apple, leaves from plants of the brassica family, lettuce leaves and assorted wild plants. One seed mixture that has been found to be most adequate consists of 80 per cent plain canary, 10 per cent black rape and 10 per cent niger – and if this is fed it is totally unnecessary to purchase expensive condition seed mixtures.

Baths, either fixed to the front of the cage or free-standing on the floor of the flights, should be offered daily, a practice that can be extended throughout the year. These should be removed at lunchtime to ensure that the birds do not take a bath late in the day and go to roost with wet feathers.

Should a bird show any sign of being distressed as evidenced by its constantly sitting around with feathers fluffed out, it should be removed from the flights and caged separately and its seed diet restricted to plain canary and maw seed in equal proportions. Should the condition not improve, or if it deteriorates to the point where the bird is sitting on one leg with its head tucked under its wing, it should be placed in a specially thermostatically controlled heated cage, the thermostat being set at about 30°C.

In places it is possible to purchase such cages, known as 'hospital cages', but it is not difficult for a breeder to manufacture a makeshift version in an emergency using electric-light bulbs as the source of heat.

A constant supply of mineral grit should also be made available if any substance other than sand is used as the floor covering. Birds do not have teeth and they need the grit as an aid to digestion. Some breeders advocate that a piece of cuttlefish bone should also be offered as a source of calcium.

Approximately eight weeks prior to the intended commencement of the breeding season, males and females should be separated. If only one flight is available it is suggested that the males be transferred to single cages, allowing the females to have the maximum amount of exercise space.

If a false environment is to be created to force the stock into breeding condition earlier than would normally occur, the heat should be switched on with the thermostat set at 10–12 °C. The lights should also be used to extend the daylight hours by a half an hour extra every second week so that by the time the pairing-up process is completed the birds are enjoying thirteen and a half to fourteen hours of light per day.

Opinions differ on whether artificial light should be given before dawn or after dusk. Should the former be preferred it is obviously better for the breeder to have an automatic time-switch installed in his breeding-room to eliminate the necessity of arising at an early hour to switch the light on. The light in this instance should not be switched off until the sun has risen sufficiently to afford maximum natural light.

Should the second method be preferred the light must be switched on either manually or automatically at least two hours before dusk and must be gradually dimmed to darkness. In this way, the birds will roost naturally whereas, if the light is suddenly turned off, they will be subject to fright and distress and will most probably hurt themselves by panic flight. It must also be remembered that if the latter method is used the female parent will have no prior warning that dusk is approaching and will not ensure that her youngsters have sufficient food in their crops to last them through the night. The breeder must, therefore, check before setting off the dimming process and if necessary hand feed the youngsters himself. This can be done by further moistening the rearing food and offering it to the chicks either with a modified hypodermic syringe (minus the needle of course) or on the end of a small soft twig.

At the time of the separation of the males and females a modified rearing food should be offered in small quantities. Special dishes that clip under the door can be used or alternatively free-standing dishes can be placed on the floor of the cage. This should initially be offered once a week. There are very many versions of rearing foods available. Most commercially produced varieties are readily accepted by the birds and prove to be suitable for rearing young. Some need additives, some do not, but in most cases the instructions for use are clearly printed on the label. Other breeders prefer to make their own mixture, but as these vary greatly, until a beginner has gained experience he would probably be better advised to use a commercially made product. Once the choice has been made, the beginner is advised not to alter either seed or rearing food mixtures until the breeding season is completed.

The selection of stock having been completed, when the birds were obtained, by the time of pairing the breeder will know which females he intends mating with which males, and in consequence all that is now needed is an explanation of the system to be used.

Prior to moving the females from the flight to their breeding cages, these cages should be thoroughly scrubbed out with hot water to which a strong disinfectant has been added. When dry, the cages, whether doubles or trebles should be divided by means of slides into single units. If it is intended to mate the male with two females he should be placed in the centre cage with the females on either side. The slides used should either be of wire construction, similar to a cage front, or if of wood should have a number of holes of about 20 millimetres diameter drilled into them about 50 millimetres from the bottom. This enables the partners to see each other and allows the male to feed the female but prevents any other contact. This is important because if the birds are put together and are not totally in breeding condition they will often fight ferociously, and it has even been known for one of a pair to kill its intended mate.

If the birds are fit, normally after a period of one week, the male can be observed feeding his mate(s) through the holes of the slide and it is then generally safe for one of the slides to be withdrawn. If the male is in peak breeding condition he will be observed drawing in his feathers tightly to his body and singing lustily. The female will be observed calling to her mate and carrying any loose pieces of paper, wood shavings, etc. in her beak in an urgent manner and also, when the male is singing she will squat on the perch with her tail raised. This is a good indication that both members of the pair have accepted each other and on meeting copulation will take place.

If two females are to be mated to one male he should be introduced to them each in rotation. Both slides should not be withdrawn at the same time as all three birds must not be allowed to be together. Females in breeding condition normally resent the presence of another female and will fight. Having seen copulation completed, some breeders withdraw the male immediately, and then introduce him to the second female, four or five hours later. Others prefer to leave the male with the first female for half a day, and then transfer him to the second female for the rest of the day. In cases where it is impossible for the breeder to effect this change in the middle of the day, it is suggested that the male be transferred from one female to the other on a daily basis. The life of the pairs should continue in this manner for a further week and then the nest-pans can be introduced.

Nest-pans are manufactured in a number of forms, those made of plastic, wicker and earthenware being most commonly used. If either of the last two are used, these should have been soaked in a solution of an insect repellent and then thoroughly dried a week or two before use. Nest-pans can be fitted anywhere in the cage so long as they are easily accessible and are fixed so that they will not fall. The plastic versions are fitted with a hole through which a screw can be fixed to hold it in position, while the wicker and earthenware types are provided with a special wire holder. Some people prefer to adapt their nest-pans so that they hang on the front of the cage. With the all-metal cages mentioned earlier, a special nest-pan attachment can be fitted on the outside of the cage. These are particularly useful in that, firstly, the nest-pan is not taking up space in the cage, and secondly, inspection of the nest and its contents is so much easier.

The material used to make plastic and earthenware nest-pans, is of a shiny smooth nature and needs a lining of felt or baize so that the female, when making her nest, can firmly attach the chosen nesting material. As nest-pans usually have small holes drilled in the bottom, the lining can be either sewn or stuck into position. Almost any glue is suitable but it is preferable to melt down a bar of carbolic soap and use this instead. A minimum of two nest-pans per pair is necessary as will become obvious later.

A canary will normally build her nest with almost any material offered, but it is customary for mosses, soft dried grasses and small pieces of soft string to be used. These can be suspended between the bars of the cage front or placed in special wire containers. Occasionally a female will reject all nesting materials, preferring instead to use paper or wood shavings. When this happens, the fancier must attempt to build a nest himself with the usual materials using something like a small electric-light bulb to compress them into place. Once the female has laid her second egg the nests are exchanged and it is almost unknown for a hen to not accept this substitution.

When the site of the nest-pan has been decided some thought should be given to ensure that, when the female has laid her clutch of eggs and is incubating them, she will not be in a position where she will be exposed to direct rays of sunlight. If this is unavoidable, it is recommended that the windows are painted with a diluted solution of white emulsion paint. This will not greatly reduce the amount of light entering the building, but will prevent direct sunlight from disturbing the birds. It can also be readily removed once the breeding season is completed.

A female canary will lay four eggs in an average clutch. As they are laid it is recommended that they are removed and replaced with dummy substitutes. Care must be taken to store them in a site that is not subject to extremes of the weather and also to label them clearly so that it is known from which pair they originated. These can then be returned to the nest on the morning that the female lays her fourth egg. The incubation period for canaries' eggs is fourteen days, although it is not uncommon for youngsters to hatch a day or two before or after this. If the eggs are initially removed and then replaced on the fourth morning, all of them may be expected to hatch on the same day. Having replaced the eggs, a record should be made of the expected hatching date. Some females start to incubate after laying their first egg, and if the eggs are left in the nest the first chick to emerge will be four days older than the last. Although extremely tiny on hatching, baby canaries grow rapidly and the growth rate in four days is considerable. This being so, the larger chick, because of its size and strength, is likely to receive more food from its mother than its smaller nest mates, which often results in the smaller birds literally being starved to death.

Occasionally, particularly if a female is not fully fit, she will be unable to pass her egg. This condition is known as 'egg binding' and the symptoms are that in the morning the female huddles in the corner of the cage or on the nest in a distressed state no egg having been laid, but when an egg could reasonably be expected. The cure is to coat the area of the vent of the bird with a warm solution of olive oil and place her in a hospital cage with the thermostat set at about 30°C. After an hour or so in this environment the egg will normally be passed successfully. Sometimes the female will go on to complete her clutch with no further problems, but equally often no more eggs will appear. If this happens it is advisable to withdraw the nest-pan, and to leave the female alone for two or three weeks and then to start the process again.

Once the clutch has been returned and the female starts her incubation (this is termed 'setting the clutch') the male is usually withdrawn totally.

When seven days have elapsed the nest can be removed and the eggs tested for fertility. This is done by holding the egg up to a bright light. If not fertile the yolk will be clearly visible rising to the top of the egg, if fertile the egg will appear dark throughout. The nest should be quickly returned if the eggs are fertile, but withdrawn totally if the clutch is clear (non-fertile). In this instance the female will usually show signs of wanting to build another nest about a week to ten days' later, and the whole process recommences.

Assuming, however, that all is well and the chicks hatch on schedule, a constant supply of rearing food should be offered. The only necessary additive to the mixture that was offered when the birds were being prepared for the breeding season is hard-boiled eggs, well mashed. Soaked seeds and chopped lettuce or chicory are also often either added to the mixture, or offered separately. Many fanciers also offer seeded weed heads (chickweed and dandelion being the favourites) but the increased use of chemical pesticides in agricultural areas makes this a progressively dangerous gamble.

When the first egg-shells are noticed on the floor of the cage announcing the arrival of the first chicks, the nests should be checked to ensure that, firstly, the broken half of one shell has not slipped over the end of another egg and, secondly, that a chick, having started to emerge from an egg has not for some reason failed to completely crack open the shell. The remedies for both mishaps are quite simple. In the first instance the whole egg should be put in a shallow dish of water whereupon the loose shell will easily separate from the whole egg and in the second instance the shell should be gently eased open continuing from the area already cracked.

The essence of stock improvement and the accumulation of knowledge is based mainly on the observation of one's own stock. The use of the memory to record all these observations for future use is haphazard, and it cannot be emphasised too strongly that accurate and complete records are absolutely essential. Observations and records should start from the day that the bird is hatched and should be continued until it has completed its moult. Particular attention should be given to anything that could be of relevance in the future. This includes the colour of the skin, legs, beak and feet, nest feather as it forms, and both the lipochrome colouring and the deposition of melanistic pigment. Similar observations should be made of the adult plumage.

During the first six or seven days the female will keep her nest clean by eating the droppings passed by the chicks. These droppings are contained in a faecal sac and when the chick is healthy this sac will remain intact. Should the chick have some form of internal disorder the sac will burst when being passed, making it impossible for the female to keep the nest clean. The droppings are thus deposited round the nest and eventually transferred on to the plumage of the mother making it appear that she is sweating. Should this occur, all rearing foods and green food should be withdrawn and substituted with a mixture of stale brown bread and boiled milk that has been cooled to an acceptable temperature. To make the mixture more appetising a small quantity of glucose and maw seed may be added. The probable cause of such an upset is that the rearing mixture is too rich for the chicks to digest properly and the bread and milk mixture will usually correct this problem.

Once the chicks are about a week old, the female will stop cleaning the nest and it is at this time that the youngsters should be fitted with a closed ring on which is stamped an identifying number and the year. In some countries it is compulsory for closed rings to be fitted before a bird can be exhibited. In Great Britain this is not so, but the practice is recommended as it affords a permanent means of identification. The bird's inheritance can always be recalled providing details were properly recorded at the time the ring was fitted.

The method of fixing such rings is as shown in the diagram below. The bird should be firmly held in one hand with the chosen leg held by the thumb and first and second fingers. The ring should be placed over the three front toes initially with the back toe pulled back along the leg. The ring should then be pushed as far up the leg as is necessary to clear the back toe and then released. Although the age of six to eight days is suggested as being the correct time to fit closed rings it could be either too soon or too late. Canaries do not increase in size uniformly, some remain comparatively small and when the ring is fitted it will slip off. Others grow very fast and at the suggested age the thick part of the foot is too large for the ring to pass over easily. Should the former be the case, attempts should be made later until the ring remains in place. If the latter, the foot should be lubricated with vaseline or water and then a firm but gentle attempt made to pass the ring over the thick part of the foot. If it is obvious that only by applying great pressure will it be possible to fix the ring, the attempt should be abandoned. Only by experience will the breeder be able to judge at what time his birds need to be ringed.

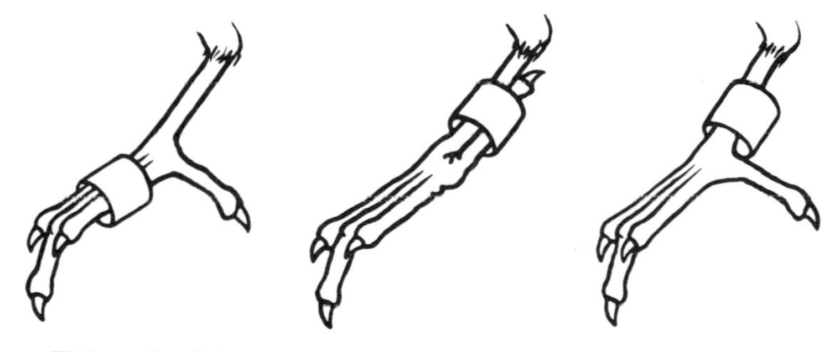

Fitting a closed ring

If the rings are fitted before the female has finished cleaning the nest, it is possible that she will succeed in ejecting the offending ring from the nest – with the chick attached. It is better to ring the birds late in the evening when the light is starting to fade and the mother is less likely to spot the rings. Even so it is always necessary to check after the birds have settled for the night that no chick has been ejected.

In the event of a chick being found on the floor next morning, even if it appears cold and dead, all need not be lost. The chick should be cupped in the hand and warmed by the breeder exhaling warm breath on to it. Unless the bird has been on the floor for a considerable length of time, it will usually revive and can then be replaced in the nest.

When the chicks are fourteen or fifteen days old, they will be almost fully feathered and the mother will probably stop brooding them. At this time the nest-pan should be placed on the floor and a new one fitted in the original site. Nesting materials should be offered as with the first nest and the female, after a day or two, will usually start to build a second nest, while at the same time continuing to rear her first brood. The male should then be reintroduced and can either be left with the family or removed immediately after copulation has taken place, depending upon whether one or two females are being serviced.

By the time the female is nearing the completion of her second nest, the youngsters will have become adventurous and will be starting to leave the nest and to feed themselves. At this time, should the female pluck the feathers from them to line her second nest, or should the young become a nuisance by deserting their first nest and claiming the new one, they should be separated by replacing the slides which were used earlier when the male was first introduced to the female. The chicks should, of course, be put on the opposite side of the partition from the nest and their mother who will then feed them through the holes in the slide.

Initially small amounts of rearing food should be offered to the youngsters in a separate dish and, if not eaten, should be replaced with a fresh supply at least twice a day. It is normal for the youngsters to be seen feeding themselves prior to the female's second clutch being set, but, if they are not, there is no need to worry. The female will continue to feed them as required until they are self-sufficient while also incubating her second clutch.

The Moult

Immediately the youngsters are self-sufficient they should be removed to a separate cage in which perches have been set up at varying heights.

Sometimes the back toe will not grip the perch properly and it will project forwards with the three front toes. This is known as 'undershot claw'. In other cases the back toe will lay back along the leg, we call this 'slip claw'. The treatment that needs to be effected

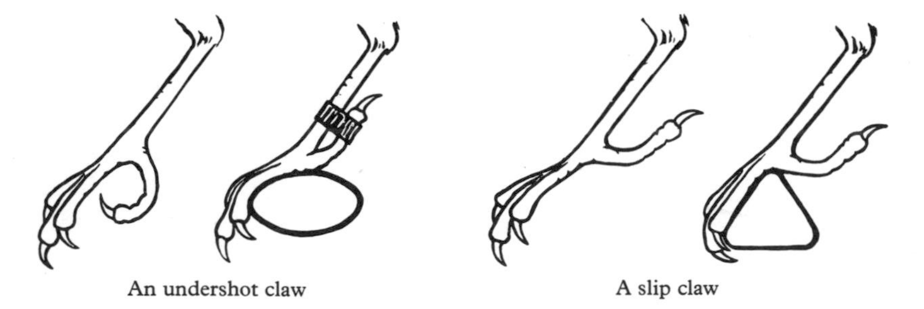

An undershot claw　　　　A slip claw

with these problems is as shown in the diagram. If the claw is undershot it should be pulled back and taped to the leg with adhesive tape. If the bird has a slipped claw very thin or triangular section perches should be installed to force the bird to grip with all claws in order to maintain its balance.

No seed should be given to the chicks initially but ample quantities of rearing food should be constantly available. When the birds are six to seven weeks old a dish containing seed can be offered, with the final choice of mixture being decided upon by the breeder. Traditionally, particularly with the red orange or rose ground birds the seed mixture will consist of niger and pinhead oatmeal or groats, while other breeders feed the seed mixture as offered to adult birds. It will be observed that when the young reach this age the tail and wing feathers are fully grown. This is a sign that they will soon commence their first moult.

During this moult they will replace all their body feathers but not the long wing and tail feathers, and if it is intended to colour feed the birds, it is now that it should commence. This only applies to red orange and rose ground birds and if the stud is mixed to include white and yellow ground birds they must be segregated before colour feeding commences.

Manufactured colour food can usually be offered in the drinking-water or mixed into the rearing food, but this is dependent upon whether the product is oil or water based.

In Great Britain, the most commonly used substance is a product called 'carophyll red', which is water-soluble and can, therefore, be offered in either way. The author finds it easier to mix the agent into his rearing food using a heaped teaspoonful (about 3–4 grammes) to 1·5 kilogrammes of the rearing food in its dry state. By virtue of its strength it is only necessary to use small quantities and if alternative systems are used whereby the agent is dissolved and used either in the drinking-water or as a moistener for the rearing food, it is difficult, if not impossible, to ensure that the same strength of solution is mixed daily. It is far easier to ensure that the same strength of solution is offered daily if a comparatively large amount is mixed with the rearing food in its dry state. The colouring agent should be given every day until the bird has finished its moult, and then offered once a week until the show season is completed. This ensures that any feathers accidentally lost after the moult will, on growing again, attain the same colour as the one lost, which is essential if the bird is to be exhibited.

If required, as an additional source of colouring, grated or whole carrots can also be fed. During the moult, baths should be offered daily and any bird that shows any of the characteristics required in an exhibition specimen should be caged by itself to avoid the possibility of another bird pecking feathers from it. This is not too important where the body feathers are concerned but if tail or wing feathers are lost, their replacements will be of different colour. This would give an unbalanced effect as the unmoulted feathers will still appear in their natural state.

As the weather becomes warmer, in spite of all precautions, a parasite known as 'red mite' may suddenly appear on the birds' bodies. This pest, although not fatal, weakens the birds, firstly, by its action as an irritant and, secondly, because it lives on the blood sucked from its victim. Should this pest be encountered, all birds and cages must be thoroughly sprayed with a special solution manufactured for the purpose and the process repeated until no signs of it remain.

The same principles should be adopted with the second-round youngsters as were used in the first round.

Some breeders will allow the females to have three rounds but this is asking a lot, particularly if the female has already raised two broods. If she shows signs of wanting to build a third nest, facilities should be provided to allow her to do so but, when she has completed laying her clutch and has sat for two or three days, the nest-pan should be removed and she should be placed in the flight with the youngsters.

Normally, adult birds will start to moult in midsummer and will finish in early autumn. The males can also be placed in the flight when their services are no longer required.

Show-training

When the moult is nearing completion, those youngsters suitable for exhibition will have become increasingly evident and they should be separated for training. It is to their advantage if they are steady and sit correctly in the showcage when being judged, thus enabling their good points to be properly evaluated. The training is best started by fixing a showcage to the front of the breeding-cage, so that the bird has ready access to both cages. At first it may be necessary to put some titbit in the showcage to encourage the bird to enter. Within a short time, however, it is usually found that it will happily hop in and around both cages.

After a week or two the bird will have sufficient confidence to be shut into the showcage. At first this should be for a short period, gradually increasing the time spent in the cage until the bird shows no form of fright or distress however long it is left there.

Although visitors should be politely discouraged from entering the breeding-room when incubation and rearing is taking place, they should be encouraged during show-training so that the birds become used to being viewed at close quarters by a variety of persons. Birds that have never encountered, for example, people with spectacles or hats or even other birds with a white ground colour will sometimes show panic on confrontation. It is, therefore advisable that they are subjected to as many different experiences as possible before being sent to an exhibition.

With the moult completed the breeder is able to review the results of his efforts and the difficult task of stock selection begins again. Certain of the young will show the points for which he has been searching, others will be disappointing. If the records that have been kept are accurate, they should help him to decide which birds should be retained for the next breeding season and which should be disposed of. Defects of a specific nature within the stud should also be noted so that new stock can be introduced to correct the fault.

The Show Season

At the start of the show season, the breeder should try to select a minimum of two teams so that if he intends to exhibit each week he can alternate his teams, thereby not unduly tiring any bird.

Great care should be taken when reading the schedule of a show before completing the entry form in order to ensure that the bird is entered in the correct class. If he has any doubts whatever, the beginner should ask an experienced breeder for advice. There is no greater disappointment than to arrive at a show full of expectation and then to find that the judge has disqualified his bird.

When visiting a show after judging has taken place, if his birds have not won their classes, he should study those that have beaten them very closely to see in what way they are superior. If this is not obvious the advice of the judge or an experienced breeder should be sought and an effort made to discover the type of stock to be introduced into the stud to attain greater perfection in the following year.

In spite of all the frustrations encountered, when the day does arrive on which a bird that has been bred is acclaimed the best exhibit in its class, then the feeling of personal pride will be indescribable.

Coloured Canary Nomenclature

Lipochrome Varieties

Ground Colour	Red Orange	Rose	Ivory Silver	Ivory Gold	Recessive White
Non-Frosted	Red Orange	Rose	Ivory Silver	Ivory Gold	Recessive White
Frosted	Apricot	Rose	Ivory Silver	Ivory Gold	Recessive White

With the exception of red orange ground birds, lipochrome varieties are qualified by the addition of the words 'non-frosted', 'frosted', or 'dimorphic' to indicate feather type.

Self Varieties

Ground Colour *Classic Variety*	*Brown*	*Green*	*Isabel*	*Agate*
Red Orange	Red Orange Brown	Bronze	Red Orange Isabel	Red Orange Agate
Rose	Rose Brown	Rose Bronze	Rose Isabel	Rose Agate
Gold	Gold Brown	Green	Gold Isabel	Agate
Ivory Gold	Ivory Gold Brown	Ivory Green	Ivory Gold Isabel	Ivory Agate
Silver (Dominant White)	Silver Brown	Blue	Silver Isabel	Silver Agate
Ivory Silver	Ivory Silver Brown	Ivory Blue	Ivory Silver Isabel	Ivory Silver Agate
Recessive White	Recessive White Brown	Recessive Blue	Recessive White Isabel	Recessive White Agate

The terminology used to describe all self varieties follows a set pattern. Initially the feather type is quoted. This is followed by the ground colour and the classic variety involved. Lastly the mutation is named. Where there are more than one, they are listed in order of appearance. Thus an isabel opal with a mutated yellow ground colour and a frosted feather is listed as a frosted ivory gold isabel opal.

Glossary

Allelomorph One of a pair of alternative heritable characters.

Body cell The unit of living matter of which animal tissues are composed.

Cell A unit consisting of nucleus and protoplasm which compose the bodies of plants and animals.

Chromosomes Bodies present in the cell upon which are borne the genes.

Clear A bird totally devoid of melanistic pigment.

Dimorphism The condition of having two different forms.

Dominant character When, on crossing two true breeding individuals that show contrasting characters, all the young exhibit the character of one parent. This character is called the 'dominant character'.

Eumelanin black The black melanin located down the centre of the feather and on the underfeather in green series birds.

Eumelanin brown The brown melanin located down the centre of the feather and on the underfeather in brown series birds.

F Symbol for filial generation.

F1 First filial generation. The young produced from a first cross.

F2 Second filial generation. The young produced from two F1 individuals.

Factor *See* Gene.

Fertile Able to produce functional germ cells.

Fertilisation Union of a male gamete with a female gamete to form a zygote.

Flighted A bird over one year old.

Foul A self bird with feathers in wing or tail that are devoid of dark pigment.

Frosted Having the extreme tip of a feather unpigmented.

Gamete *See* Reproductive cell.

Gene A particle of substance situated on a chromosome in the germ cell which is responsible for the expression of a given hereditary character.

Genotype The genetic constitution of an individual. A group of individuals genetically identical.

Heredity The factor in evolution which causes the persistence of characters in successive generations.

Heterozygote A bird which carries both members of an alternative pair of genes, hence a bird which cannot breed true to either of the two characters involved.

Heterozygous Possessing both the dominant and the recessive genes of an alternative pair.

Homozygote A pure bred. A bird which must breed true to a specific character, as it carries in duplicate only one member of an alternative pair.

Hybrid The offspring of two different species.

Inheritance That which is or may be inherited.

Lipochrome Fat-soluble feather colouring material.

Melanins Black and brown pigments on self birds, formed from protein produced by the birds.

Mendelian character A character which is inherited according to 'Mendel's Law'.

Mutation A spontaneous change in the constitution of a gene.

Non-frosted A feather on which the lipochrome extends to the tip.

Ovary The female reproductive gland producing ova.

Phaeomelanin brown The brown melanin located on the edges and tip of the feather in self birds.

Phenotype The sum total of the hereditary characters apparent in an individual. A group of individuals all of which look alike.

Recessive character Of a pair of allelomorphic characters, the one which will not be manifested in the young if the genes for both characters are present.

Reproduction The process whereby life is continued from generation to generation.

Reproductive cell The gamete or germ cell. The spermatozoon produced by the male and the ova produced by the female.

Segregation The separation in a heterozygote of the two members of a pair of allelomorphic genes.

Self A bird that has pigmentation in all its feathers.

Sex chromosomes The chromosomes in respect of which the male and female differ.

Sex linkage Association of a hereditary character with sex as its gene is situated on the sex chromosome.

Spermatozoon or *Sperm* The male germ cell.

Sterile Unable to breed.

Ticked A bird with one dark mark that can be covered by a British 1p piece.

Unflighted A current year bred bird.

Variegated A bird with more melanistic pigment than a ticked specimen, but also has some areas of lipochrome feathering visible.

X Chromosome The male sex chromosome.

Y Chromosome The female sex chromosome.

Zygote The single cell formed by the union of a male and a female gamete.

Index

Bold figures refer to colour pages.

Societies

The Canary Colour Breeders Association is a specialist society that caters expressly for the needs of Coloured Canary Breeders.

It does this by staging its own exhibition as well as promoting other exhibitions throughout the country.

Within the CCBA are zonal societies that assist breeders in their locality, a Genetical Section, whose members conduct group experiments in an attempt to unravel some of the outstanding mysteries associated with Coloured Canaries and a Judges Convention where the constantly changing show standards are debated.

Two magazines are distributed annually to its international membership.

Full details can be obtained from the Secretaries:

Mr and Mrs John Pitcher,
16 Detling Avenue,
Broadstairs, Kent

The Confederation Ornithologique Mondial is a non-profit-making organisation, the membership of which is open to every country in the world. Only one organisation from each country can be affiliated and the British Society is the International Ornithological Association of Great Britain.

The objects of the COM are:

(*a*) The spread of knowledge of ornithology throughout the world in order to arouse interest in, and to protect, birds.

(*b*) To promote and encourage all matters relating to birds and those interested in birds.

(*c*) To develop a friendly understanding and spirit of co-operation among the countries of the world, and between all bird breeders and fanciers.

(*d*) To maintain a permanent Panel of International Judges.

(*e*) To study existing standards and, where necessary, to draw up exhibition standards for all species of show birds in order to evolve a single international standard.

(*f*) To foster the exchange of birds between members of the COM.

(*g*) To organise exhibitions and international contests.

(*h*) To exchange ornithological periodicals and technical publications.

The Panel of Judges mentioned in paragraph (*d*) having passed both practical and written examinations are admitted as members of the Ordre Mondial des Juges, their own very special society within the COM.

It is from this society that judges are selected to officiate at the World Championship Shows, which are staged by a different country each year.

The Secretary of the IOA is:

Chas. H. Smith, F.Z.S.,
7 Cleve Road,
Sidcup, Kent, DA14 4RS